T0184698

S3D Dashboard

Florian Weidner

S3D Dashboard

Exploring Depth on Large
Interactive Dashboards

Florian Weidner
Ilmenau University of Technology
Ilmenau, Germany

PhD Thesis, Technische Universität Ilmenau, 2021

ISBN 978-3-658-35146-5   ISBN 978-3-658-35147-2   (eBook)
https://doi.org/10.1007/978-3-658-35147-2

Responsible Editor: Stefanie Eggert
This Springer Vieweg imprint is published by the registered company Springer Fachmedien
Wiesbaden GmbH part of Springer Nature.
The registered company address is: Abraham-Lincoln-Str. 46, 65189 Wiesbaden, Germany

"After all, you can't have the applications without the curiosity-driven research behind it."

Prof. Dr. Donna Strickland
Recipient of the Nobel Prize in Physics 2018.

Acknowledgements

When I started this endeavor years ago, I did not really know what to expect. Luckily, I had the pleasure to work with people and in places that made this journey an enjoyable one. I want to highlight and thank some of them in the following—those who had the biggest impact during this journey.

First and foremost, I want to thank the department of Computer Science and Automation of TU Ilmenau for giving me the opportunity to write this dissertation. I also want to thank the committee members and especially the reviewers of this thesis for their time and effort. I really appreciate it!

Further, I want to thank my supervisor, Prof. Wolfgang Broll, who gave me all the freedom I could wish for. You always provided the space (and equipment) for my creativity and for my scientific development. You always trusted me and encouraged my independence. Thank you for everything. I also want to thank Anja Hofmann who always had an open ear and helped me with all the administrative work.

To my colleagues and fellow graduate students Tobias Schwandt, Christian Kunert, and Christoph Gerhardt—thank you for your feedback, for the fruitful discussions, for the countless breaks and talks. Working with you has been a wonderful experience and I had the pleasure to learn a lot from you during this time. I wish you all the best for your future endeavors! To all the other people I crossed path with and especially to Anne Hösch, Sandra Pöschl, Bernhard Fiedler, and Johannes Redlich: thank you for the work we did together but also for satisfying my curiosity by giving me insights in your research. I hope our paths cross again in the future.

I also had the great pleasure to supervise some outstanding students during the time of this thesis. Their work influenced this thesis and they supported me with it.

More importantly, their spirit and curiosity encouraged me and constantly reminded me, why research matters. My special thanks go to Aliya Andrich, Kathrin Knutzen, Christine Haupt, and Luis-Alejandro Rojas-Vargas for their great work. It was my pleasure to work with you and I wish you all the best.

To all of my friends who had to endure me during this time, especially during the more stressful episodes: thank you for always being there for me. Without your encouragement and support, I could not have done it. And finally, to my family—my mum, my dad, and my brother. I'm really lucky to have you in my corner. You always had my back and supported me in any way you could. It really means a lot to me. I can't express my gratitude in words but believe me, it is limitless.

To everyone else who supported me during this journey, thank you!

Abstract

Over the last decades, the interior of cars has been constantly changing. While the fundamental interaction devices—steering wheel and pedals—stayed the same, other parts evolved. Adapting to developments like driving automation, novel in- and output devices have been integrated into the vehicle. A promising, yet unexplored, modality are large stereoscopic 3D (S3D) dashboards. Replacing the traditional car dashboard with a large display and applying binocular depth cues, such a user interface (UI) could provide novel possibilities for research and industry. To enable research on S3D dashboards, we introduce a development environment based on a virtual reality environment simulation and a car mock-up featuring a spatial augmented reality dashboard. Using this environment, we define a zone of comfort for viewing S3D content on dashboards without the loss of binocular fusion. With these results, we performed several driving simulator experiments, covering use cases from manual as well as automated driving. Considering the prior evolution of in-car UIs and the advances in vehicular technology, the focus lies especially on driver distraction while driving manually, trust, take-over maneuvers, and non-driving-related tasks. We show that S3D can be used across the dashboard to support menu navigation and to highlight elements without impairing driving performance. We demonstrate that S3D has the potential to promote safe driving when used in combination with virtual agents during conditional automated driving. Further, we present results indicating that S3D navigational cues improve take-over maneuvers in conditional automated vehicles. Finally, investigating the domain of highly automated driving, we studied how users would interact with and manipulate S3D content on such dashboards and present a user-defined gesture set. Taking the core findings into consideration, we outline the strengths and weaknesses of this UI technology.

In the end, this work lays the foundation for future research in the area of large interactive S3D dashboards by presenting design and development tools, results of user studies, and possibilities for interaction metaphors. Findings act as basis but also motivation for further research, indicating that large interactive stereoscopic 3D dashboards can play a viable role in future vehicles.

Kurzfassung

Die Benutzungsschnittstelle von Fahrzeugen hat sich in den letzten Jahren ständig verändert. Die primären Interaktionsgeräte—Lenkrad und Pedale—blieben nahezu unverändert. Jedoch haben Entwicklungen wie das automatisierte Fahren dazu geführt, dass neue Ein- und Ausgabegeräte Einzug hielten. Eine vielversprechende, jedoch kaum erforschte Modalität sind stereoskopische 3D (S3D) Armaturenbretter. Das Ersetzen des traditionellen Armaturenbrettes mit einem großen Display, welches binokulare Tiefeninformationen darstellt, könnte neue Möglichkeiten für Forschung und Industrie bieten.

Um Forschung an dieser Technologie zu ermöglichen, stellen wir eine Entwicklungsumgebung. Diese besteht aus einer Virtual-Reality-Umgebungssimulation und einem Fahrzeug-Mock-up. Das Mockup ist mit einem Spatial-Augmented-Reality-Armaturenbrett ausgestattet. Weiterhin wurde für dieses Setup eine Zone für komfortables Sehen von S3D-Inhalten ermittelt. Darauf aufbauend wurden Anwendungsfälle aus den Bereichen des manuellen und automatisierten Fahrens untersucht. Basierend auf der Entwicklung von Fahrzeug-Benutzungsschnittstellen wurde der Fokus auf Fahrerablenkung während des manuellen Fahrens, auf Vertrauen, der Wiederaufnahme der Fahraufgabe und dem Durchführen von nicht-fahrbezogenen Aufgaben während dem bedingten bzw. hochautomatisierten Fahren gelegt. Die Ergebnisse zeigen, dass S3D auf dem ganzen Armaturenbrett für die Menünavigation und zum Hervorheben von Inhalten genutzt werden kann, ohne die Fahrperformance zu beeinträchtigen. Weiterhin wird gezeigt, dass ein intelligenter Agent, dargestellt in S3D, sicheres Fahren fördert. Darüber hinaus wird gezeigt, dass stereoskopische Navigationshinweise das Wiederaufnehmen der Fahraufgabe in einem hochautomatisierten Fahrzeug verbessern. Schlussendlich wird untersucht, wie Menschen mit stereoskopischen Inhalten, dargestellt auf einem S3D-Armaturenbrett, interagieren und ein benutzerdefiniertes Gestenset für die Interaktion vorgestellt. Abschließend werden, basierend auf den Resultaten dieser Studien, die Stärke und Schwächen dieser Benutzungsschnittstelle diskutiert.

Die Ergebnisse dieser Arbeit—Entwicklungswerkzeuge, Daten der Nutzerstudien und Interaktionstechniken—legen die Grundlage für zukünftige Forschung an S3D-Armaturenbrettern. Sie sind Basis und Motivation für weitere Arbeiten und legen nahe, dass S3D-Armaturenbretter eine tragfähige Rolle in zukünftigen Fahrzeugmodellen spielen können.

Contents

Abbreviations

ADAS	Advanced driving assistant system
ASD	Autostereoscopic display
DOF	Degree of freedom
hFOV	horizontal field of view
HUD	Head-up display
IPD	Inter-pupillary distance
MBDC	Mean blink duration change
MPDC	Mean pupil diameter change
MPDCR	Mean pupil diameter change rate
NDRT	Non driving related task
P3D	Perspective 3D
S3D	Stereoscopic 3D
SA	Situation awareness
SAE	Society of Automotive Engineers
SAR	Spatial augmented reality
SSQ	Simulator sickness questionnaire
TOR	Take over request
UE4	Unreal Engine 4
UEQ	User experience questionnaire
UI	User interface
UX	User experience
VR-HMD	Virtual reality head-mounted display

List of Figures

List of Tables

Part I
About This Thesis

Introduction   1

1.1   Motivation

Since the invention of the first gasoline automobile in the 1880s[1], the way how users operate them has been constantly changing. Figure 1.1 highlights this development, showing a plain vehicle with just primary control elements next to later generations of vehicle interiors. While the core elements of the user interface (UI) — steering wheel and pedals — seem to be fixed nowadays, the rest of the controls changed, got more complex, and evolved. From having initially barely any secondary controls at all, today's cars often feature sophisticated input and output devices. For example, the Tesla Model 3 (c.f. Figure 1.1c) offers few mechanical buttons but a rather large customizable touchscreen with high-quality graphics and a complex menu structure. Other prevailing input devices are knobs and levers, rotary controllers, voice control, and mid-air gestures. Today's cars are also equipped with various output devices like analog gauges, various digital screens, and head-up displays (HUD). These advanced modalities have been necessary for various reasons: advanced driving assistance systems (ADAS) have been introduced to cars and offer customization options like special driving modes (e.g. economic driving) or different settings of the suspension (e.g. sport-mode). Also, more and more comfort and entertainment functions have been integrated into cars. For example, air conditioning and multimedia systems including radio as well as screens for passengers to watch movies. Digitization and mobile connectivity further add to the host of functions. For example, in many cars, the phone can be connected to the on-board computer to simply forward notifications or even to be used as an extension of the UI, e.g. for navigation or calling. It can

---

[1] www.britannica.com/technology/automobile/History-of-the-automobile, History of the automobile; 2020-09-23

© The Author(s), under exclusive license to Springer Fachmedien Wiesbaden GmbH, part of Springer Nature 2021
F. Weidner, *S3D Dashboard*,
https://doi.org/10.1007/978-3-658-35147-2_1

be argued that this development will continue, and that the nature of in-car user interfaces will evolve further.

(a) 1939 Graham	(b) 1996 BMW 8 Series	(c) 2017 Tesla Model 3

Figure 1.1 Development of in car user interfaces: while early cars had few controls, later user interfaces had a plethora of knobs and buttons. Today's cars show few of those but offer a lot of functions encapsulated in digital user interfaces like touchscreens

The mentioned examples show that, while driving is still the primary purpose of a car, research and industry designed the vehicle interior to be a space that offers a relaxing environment and gives access to various multimedia functions. So far, the primary design objective of automotive in-car user interfaces has been minimizing driver distraction and maximizing safety. While driving is getting indeed safer — at least according to statistics of the German Road Safety Council[2] — the problem is far from being solved, as the still high number of 3.275 fatal accidents on German roads shows. Thus, safety will still be an issue in future automotive user interface development. Here, one of the main challenges is to operate the UI safely while driving. To avoid accidents, drivers are supposed to fully concentrate on the traffic. However, they also want to operate the in-vehicle UI to control the radio, adjust the air conditioning and more. Also, they need to supervise critical information like navigation or the car's speed. This interaction distracts them from driving and is a major cause for accidents. For example, 9% of the fatal crashes in the US were related to distracted driving[3]. Hence, a good UI should support and ease interaction and reduce distraction: input devices should allow interaction that is fast, precise, intuitive, and — at best — does not require long glances at control elements. Output devices should encode information in such a way, that it can be easily and quickly perceived and understood. By that, input and output devices can contribute to a safe drive.

---

[2] https://www.dvr.de/unfallstatistik/de/aktuell/, Unfallstatistik aktuell (translation: "current accident statistics"); 2020-09-23

[3] https://crashstats.nhtsa.dot.gov/Api/Public/ViewPublication/812700, Distracted Driving in Fatal Crashes, 2017; 2020-09-23

In addition to the issue of safety, recent developments in driving automation pose another challenge for in-car UIs: conditional, highly, and fully automated vehicles require UIs that are specifically build for these driving modes. For example, if the vehicle drives by itself, it is important that the user trusts the automation system to feel comfortable and safe [Lee and See, 2004]. Low levels of trust towards the automation could prevent using the system at all or lead to misuse [Lee and See, 2004]. It has been shown that user interfaces can foster trust, e.g. by communicating system behavior [Koo et al., 2015]. Also, according to the classification of the Society of Automotive Engineers (SAE), when driving a vehicle of automation level 3, drivers need to be ready to take over the driving task at all times while the vehicle is driving by itself [SAE International, 2015]. At best, drivers always monitor the road and supervises the vehicle. By that, they are able to correctly react and take over control if necessary. However, it has been shown that humans perform subpar at monitoring tasks. They easily lose situation awareness (SA), which can lead to potentially dangerous reactions in critical situations [Endsley, 2017]. Advanced user interfaces like ambient lighting [Borojeni et al., 2016] or haptic cues [Petermeijer et al., 2017b] can support this take-over process. In addition to that, driving automation enables performing non-driving related tasks (NDRT). For example, with SAE level 4 automation, the car can drive by itself, and no supervision is necessary [SAE International, 2015]. Then, the driving task is executed completely by the vehicle and no sudden reaction is necessary. When driving such a vehicle, driver-passengers will always have an appropriate amount of time to take over the driving task [SAE International, 2015] — the process is non-critical. Hence, the user interface could be fully dedicated to NDRTs and change back to the driving mode when necessary. For example, driver-passengers of future highly automated vehicles could surf the internet in their free time [Pfleging et al., 2016], but could also be required to do certain work-related tasks, e.g. taking part in online trainings or do data processing [Frison et al., 2019; Teleki et al., 2017]. Today's vehicles have not been designed for such a switch and often hamper the ergonomic execution of non-driving related tasks.

Overall, it can be assumed that in-vehicle user interfaces will play an even more important role in the future and that they need to evolve to meet (or keep up) with the discussed challenges: besides the aspect of safety, objectives of future interface development are, among others, designing them in such a way that users' trust in automation is sufficiently well build and maintained, that they support users in critical situations when automation fails, and that they support NDRTs when the vehicle performs the driving task.

1.2   Problem Statement

It can be argued that traditional UIs cannot fulfill these requirements. On the one hand, reducing driver distraction is an integral part of automotive user interface design but it is hardly solved, as statistics show. On the other hand, today's user interfaces were not designed for the new challenges introduced by driving automation mentioned earlier. While today's cars offer a relatively safe experience, they do not necessarily foster trust in automation or support take-over processes. Also, vehicle interiors are not yet designed to allow for work or well-being related tasks without causing discomfort or pose a safety risk. In addition to that, the problem of safely issuing and handling a time-critical take-over is still unresolved. Hence, there is clearly a need for user interfaces that can meet these requirements: increase safety, support take-over, allow for non-driving related tasks, and foster trust.

A possible solution for these issues could lie in the third dimension: stereoscopic 3D (S3D). Integrating binocular depth cues into in-car displays could potentially provide novel design possibilities for research and industry. By adding another dimension to the design space, information could be presented in a way that is closer to reality. By that, it could add value to the user interface: for example, binocular depth could act as a new design element and foster understanding of navigational information. Or, if such a S3D display provides a large interactive workspace (similar to a tabletop), the in-car UI could enable users to perform various NDRTs. It would then be possible for users to perform entertainment tasks like watching a movie or to execute vocational tasks like research activities, online training, or complete course work. Previous research has dealt with two integral aspects of this approach: on the one hand, small and non-interactive stereoscopic 3D displays [Broy, 2016] and on the other hand large 2.5D interactive surfaces [Rümelin, 2014]. However, no prior work has investigated the combination of both: large stereoscopic 3D dashboards.

1.3   Objectives & Research Approach

This thesis explores large stereoscopic 3D dashboards as a possible user interface for future cars. The primary research question has been formulated as:

> *To what extend can large interactive stereoscopic 3D dashboards*
> *act as an in-car user interface for future vehicles?*

The fundamental question of this thesis is to investigate the potential of large interactive S3D dashboards in vehicles. Given the lack of prior research dealing with this problem, this thesis will derive design guidelines and tools for development but also investigate various application areas. Based on this motivation, sub-research questions were formulated as follows:

RQ1: How can large interactive stereoscopic 3D dashboards be designed and developed?
This thesis aims to provide insights on the design space of large interactive S3D dashboards. To do this, tools and methods are necessary to develop prototypes and test them in driving situations. Prior driving simulators and development tools for automotive UIs have dealt only with 2D UIs or small-scale S3D displays. Hence, we design and develop a set of tools, facilitating augmented reality and virtual reality, that allows for the development of S3D dashboard designs. It is further necessary to define a region or volume where S3D content can be displayed without leading to visual discomfort and loss of binocular fusion.

RQ2: What are viable applications for large interactive stereoscopic 3D dashboards?
To assess the potential of S3D displays, this thesis will investigate several exemplary use cases. Due to the lack of any commercially available vehicles, the exploration will be performed by means of driving simulator experiments using UI prototypes. Besides problems and challenges of today's UIs (distracted driving), especially challenges posed by recent advances in vehicle automation (take-over, trust, NDRTs) will play a role during the selection of the case study domains.

RQ3: What are appropriate interaction techniques for stereoscopic 3D dashboards?
While the interaction with 3D objects has been researched in other domains, prior research on automotive S3D displays has always investigated non-interactive use cases. Interacting with 2D or 2.5D content can be easily realized using touchscreens, but doing so with S3D content adds an additional layer of complexity due to binocular depth cues and the spatial nature of the interactive elements. The answer to this sub-research question will provide insights on how people are going to manipulate content on large interactive S3D dashboards. We intend to investigate this by means of a user-elicitation study.

1.4   Summary of Research Contributions

In summary, we present specifics on the design and implementation of the developed driving simulation environment for S3D dashboards: consisting of a VR environment simulation and a car mock-up with a spatial augmented reality (SAR) dashboard, it allows experiments on large interactive stereoscopic 3D dashboards. Using this setup, we define the zone of comfort for binocular vision on stereoscopic 3D dashboards. All 3D graphic elements should be displayed inside this volume to guarantee a comfortable viewing experience and binocular fusion.

Applying these findings, we present results suggesting that binocular depth cues can be used as a design element without impairing driving performance, e.g. when navigating through menus or for highlighting an element of the user interface. In addition to that, we show that stereoscopic 3D displays can foster anthropomorphism and that anthropomorphism and self-reported trust in the automation system are positively correlated. Continuing in the domain of conditional automation, we show that S3D displays can support the take-over process in critical situations by presenting navigational information in form of visualizations that show the surroundings. We further show that workload is not negatively influenced by such warnings and report on gaze behavior. In addition to that, results of this thesis outline how users intent to interact with content on large interactive stereoscopic 3D dashboards. Here, the interaction with abstract and application-specific content was studied. Results were used to derive a user-defined gesture set for interaction with large S3D dashboards. This set can be used for future studies that require the manipulation of stereoscopic 3D elements. We further propose additional research avenues for S3D displays:

- Applying and analyzing glasses-free displays (lightfield- or high-resolution autostereoscopic displays)
- Use-cases: communication of the surroundings, navigational systems, personalization, customization, and context-sensitive adaptation
- Communication (video chat with agents or humans)
- Real-world driving studies (or studies in moving simulators)

1.5   Readers Guide

In the following, Chapter 2 of Part I, outlines the basics regarding depth perception, S3D displays, driving simulators, and automated driving. Chapter 3 presents related work on in-car user interfaces, visual displays, and S3D UIs. It further relates prior work to our research objectives.

Following, Part II presents our findings on design and development methods of S3D dashboards. Chapter 4 presents the developed prototyping environment including VR environment simulation, car mock-up, S3D dashboard, and user interface development software. This environment has been used throughout this research. Chapter 5 introduces the derived design space for binocular fusion and stereoscopic 3D content.

Using these tools and findings, Part III presents case studies where we have applied the stereoscopic 3D dashboard. It reports on the driving simulator experiments that investigate change detection and system control to reduce driver distraction (Chapter 6), trust (Chapter 7), and take-over notifications (Chapter 8). While these studies did not allow any manipulation of S3D content, Chapter 9 dives into this area. It presents a user-elicited gesture set for the interaction with various S3D objects during a variety of tasks while driving a highly automated vehicle.

Part IV is the final part of this thesis. It provides a holistic discussion of the findings regarding the research questions presented in Chapter 10. The end of this part and also this thesis summarizes the findings, connects the common themes, and by that, outlines a research agenda (Chapter 11).

Background

2

This chapter introduces important core concepts that are necessary for an understanding of this thesis. Visual perception and depth cues are introduced, so are the foundations of stereoscopic 3D displays. To get a deeper understanding of the automotive domain, driving simulators and levels of automation are explained.

2.1 The Human Visual System & Depth Perception

The human eye allows us to navigate our world by offering a visual image of our surroundings. Objects reflect light which passes through the cornea, pupil, and lens onto the retina. Part of the retina is the fovea centralis where resolution and sharpness are best. On the retina, rods and cones are stimulated by the light and forward the information via the optical nerve to the human visual system. There, the final color image is formed. Besides colors and brightness, humans (and many animals) use visual information to make assumptions about depth.

Figure 2.1 shows a taxonomy of depth cues [Goldstein, 2014]. Depth cues can be divided into three major categories: oculomotor, monocular, and binocular. Oculomotor refers to cues that are based on the movement of the eyes or the lens, respectively. Using accommodation, the shape of the lens changes to focus on an object. Using vergence, each eye rotates to have a direct look at the object. That means that the direct line of sight goes from the object, through the lens, onto the fovea centralis. By that, the observed object is perceived sharp and clear with both eyes.

Monocular cues can be divided into pictorial and motion-produced cues. Pictorial cues are, in easy words, all depth cues, that can also be observed in a static photograph such as shadows or relative size (smaller objects are farther away). Motion-produced cues are based on the movement of objects. For example, motion parallax describes

F. Weidner, *S3D Dashboard*,
https://doi.org/10.1007/978-3-658-35147-2_2

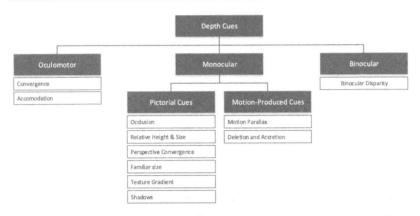

Figure 2.1 Overview of depth cues. Based on Goldstein [2014]

the change in position of the eye relative to a scene. Depth can also be determined if an objects moves in front of another object and by that occludes it (and vice versa, deletion and accretion), which is most likely the strongest depth cue [Ware, 2020].

Binocular depth cues rely on the two images provided by the two eyes. Each eye takes an image from a different perspective and captures a slightly different view of the world. For any focus state of the eyes, the reflected light of the fixated object stimulates a set of corresponding points on the retina. These sets of points build a pair of corresponding images. The location of points in space that fall on corresponding points on the retina can be roughly described by a curved plane. A vertical or horizontal line of this plane is called horopter [Howard and Rogers, 2008]. A pair of corresponding images (or a set of corresponding image points) is then fused by the human visual system—a process called *binocular fusion*. However, the human visual system shows a certain tolerance towards correspondence and does perform fusion with points that are not perfectly corresponding. That means, it also fuses points that are close to a horopter. The area in which binocular fusion is possible in general is called *Panum's fusional area* [Panum, 1858]. However, especially at the outer regions of Panum's fusional area, the human visual system experiences eye strain and discomfort while fusing the images. Here, binocular fusion is only sometimes possible and at other times, fusion is lost. This state is called *binocular rivalry* [Howard and Rogers, 2008]. The area in which binocular fusion is comfortably possible is called the *Zone of Comfort* [Shibata et al., 2011]. Factors influencing the ability of binocular fusion and the size of Panum's fusional area as well as the zone of comfort are, among others, shape, pattern, movement, color, viewing angle, and spatio-temporal resolution of the visual stimulus [Lambooij et al., 2009; Reichelt et al., 2010]. Also, both areas vary among humans. If a visual stimulus cannot be fused, humans either perceive two images (*double vision*) or the input from one eye is suppressed [Georgeson and Wallis, 2014].

The general ability to perform binocular fusion is called *stereopsis*. The inability to do so is called *binocular diplopia*. The cause for diplopia is often neurological or neuro-ophtalmological and influences the eye's movement (e.g. strabismus [Sreedhar and Menon, 2019]). Hence, not all people are able to fuse the two images of the eye. They rely on the other depth cues.

Figure 2.2 illustrates the importance of various depth cues for judging depth in accordance to the viewing distance [Goldstein, 2014]. Some cues like occlusion or relative size are important for near and far distances whereas others are only valid within short ranges, e.g. accommodation and reverence or atmospheric perspective. Binocular disparity is especially important in ranges below 20 m [Goldstein, 2014].

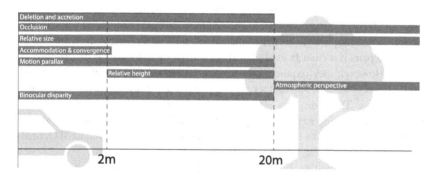

Figure 2.2 Depth cues and their effective ranges. Based Goldstein [2014]

A comprehensive overview about binocular fusion and stereopsis has been provided by Howard and Rogers [Howard and Rogers, 2008].

2.2 Stereoscopic 3D Displays

Traditional digital displays convey depth using monocular cues. They are often labeled with the terms 2D, 2.5D or perspective 3D (P3D). 2D displays rely on all monocular cues except perspective convergence whereas 2.5D displays also integrate this kind of depth cue [Naikar, 1998]. Contrary to that, stereoscopic 3D displays use special soft- and hardware to display corresponding image pairs—one image for the left eye and one image for the right eye. By that, they provide binocular depth cues.

S3D displays can be divided into two main categories: those that need dedicated glasses, and those that do not need glasses (autostereoscopic). Active 3D displays require shutter-glasses that sequentially and alternatively black out an eye so that the other one can see its dedicated image. The glasses are synced with the display, for example via radio frequency. By that, the display hardware shows the image for the eye that is currently not blacked out [Lueder, 2012]. Passive 3D systems also require glasses, but they do not use the shutter mechanism. The lenses of such glasses allow only certain parts of polarized light to pass. Hence, the display hardware must emit images with polarized light. By that, the images for the left and right eye have different polarization and can only pass through the respective glass. Alternatively, image separation can be done by wavelength multiplexing where only complementary subsets of the visible spectrum reach each of the eyes [Lueder, 2012]. According to LaViola Jr. et al. [2017], active stereo systems achieve—in general—the best stereo quality but are more expensive than passive systems. Autostereoscopic displays (ASD) do not require glasses. The display hardware is equipped with special lenses—so-called parallax barriers—that allow for the separation of the corresponding image pairs [LaViola Jr. et al., 2017]. Benefit of such displays is the absence of glasses. However, they require a compromise between image quality and viewing angle [Chen et al., 2019]. Another way to present binocular depth cues are lightfield displays. These displays also do not require glasses and promise to offer a more natural viewing experience by recreating a lightfield. Besides that, experimental technologies like holographic and volumetric displays as well as 3D retinal projection displays exist.

Spatial Augmented Reality (SAR) uses S3D displays in a special way. It allows for the presentation of S3D content using projectors that display their image on arbitrarily oriented and curved screens. These screens can be semi-transparent or opaque. In the semi-transparent case, the projector is often hidden behind the screen and usually not visible to the user. This mode is called rear-projection or back-projection. SAR makes it possible to include AR content into the naturally occurring environments [Bimber and Raskar, 2005]. We use S3D and SAR with rear-projection to build our S3D dashboard (see Chapter 4).

Besides the previously mentioned aspects that impact binocular fusion of non-artificial scenes, *binocular disparity* plays an important role in computer-generated S3D images. Binocular disparity describes the difference between the two images for the left and right eye. Figure 2.3 illustrates the concept of disparity. Here, the absolute angular disparity is calculated by subtracting the angle between the eyes and point on the screen where the object is displayed (indicated by the center cube; β) by the angle between the eyes and the location of virtual object in 3D space (indicated by the cube on the top; α). Hence, it can be defined as disparity $D = \alpha - \beta$ [McIntire et al., 2015]. A positive disparity means that the virtual content appears to be behind the screen and vice versa. A disparity value of D = 0 means that the point is on the screen. For one volumetric object, it is possible to calculate several disparity values. For example, for the cube in Figure 2.3, a front and back corner have different disparities. In this thesis, all disparity values are calculated for the geometric center, if not mentioned otherwise. Disparity also depends on the inter-pupillary distance (IPD). The IPD varies between humans. The average IPD is 63.36 mm [McIntire et al., 2015].

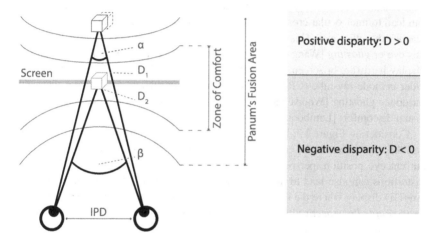

Figure 2.3 Illustration of binocular disparity, the zone of comfort, and Panum's fusional area. The cube in the upper area lies outside the fusion area and cannot be fused by the visual system. Hence, it would lead to double vision and/or visual discomfort. The center cube lies within the zone of comfort and can be comfortably perceived in S3D

If the disparity exceeds the limits of the fusion area, it can result in binocular rivalry or loss of fusion and by that, to double vision (indicated by the top cubes in Figure 2.3). However, viewing S3D content also leads to visual discomfort or visual fatigue when the content lies within the zone of comfort [Lambooij et al., 2009]. Reasons for this are manifold. Among others, conflicting depth cues, flicker, low resolution, and refresh rate or high latency of the S3D display system are known to lead to visual discomfort [Kolasinski, 1995; Lambooij et al., 2009].

Further, the accommodation-vergence conflict contributes to these symptoms. This conflict results from the fact that the eyes adjust the lenses to focus on the screen (accommodation) but the eyes also converge towards the virtual object (vergence) which can be in front or behind the physical screen. In the real world, accommodation and vergence are synchronized but S3D displays disrupt this synchronization. An ongoing mismatch leads to a visual strain and to symptoms associated with visual fatigue and discomfort [Lambooij et al., 2009]. Lightfield displays and volumetric displays promise to mitigate these issues [LaViola Jr. et al., 2017].

Technologies which rely on special glasses (active or passive stereo) might introduce additional discomfort due to their form factor but also because they impact visual perception (e.g. by limiting color space or by limiting the field of view) [Shibata et al., 2011]. Further, glasses in spatially or temporally multiplexed systems can lead to inter-ocular crosstalk where the light of an image dedicated to one eye reaches the other eye. This is also called *leakage*. The result is a ghost image in this eye or *ghosting* [Wang et al., 2011]. Another reason for ghosting is the actual display hardware or screen. Some hardware, like liquid crystal displays (LCD) or older cathode-ray-tube (CRT) screens, leave an afterimage or afterglow that can introduce ghosting [Woods and Sehic, 2009]. Ghosting is also another source for visual discomfort [Lambooij et al., 2009].

Considering Figure 2.3, it is important to note that in the real world, the head and the eyes move. If the computer-generated image is not updated to reflect the current eye position, perspective distortions are visible [McKenna, 1992]. Such distortions can also lead to increased visual fatigue [Kongsilp and Dailey, 2017]. An S3D display where the image is rendered without perspective distortion uses a *head-coupled perspective* by applying an *off-axis projection*. By using an off-axis projection, objects are visualized in a static position, regardless of user position and without distortion. This can lead to a better visualization of computer-generated content [Arsenault and Ware, 2004]. In order to apply this technique, head-tracking is necessary to estimate the position of the eyes relative to the display.

2.3   Driving Simulation

Driving simulators have a long history in research and industry. They can offer a controllable, reproducible, safe, and rather easy-to-use environment to test and develop new hard- and software [Fisher and Rizzo, 2011]. At the same time, they do not provide absolute validity. However, they show sufficiently good behavioral validity regarding driving performance measures [Mullen et al., 2011] and for the evaluation of in-vehicle interfaces [Wang et al., 2010].

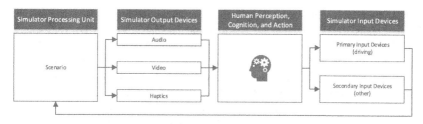

Figure 2.4 The four core components of a driving simulator. The processing unit calculates the current state of the simulation and displays it via the output devices. The human perceives, processes, and acts using the input devices which again, updates the simulation

Core components of every driving simulator are illustrated in Figure 2.4. The simulator processing unit, usually one or more computers, calculates the simulation including vehicle physics, traffic, and environment. It also controls the simulator output devices for audio, video, and haptic displays. Humans continually perceive these cues. Based on that, they perform necessary actions via the simulator input devices. Here, primary input devices are used to control the vehicles lateral and longitudinal movements whereas secondary input devices are necessary to perform other tasks (e.g. controlling a navigational system).

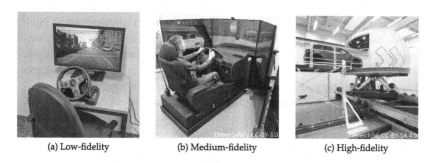

(a) Low-fidelity   (b) Medium-fidelity   (c) High-fidelity

Figure 2.5 Examples of a low-, medium, and high-fidelity driving simulators

Mainly based on the simulator output devices, driving simulators can be grouped into low-fidelity, moderate-fidelity, and high-fidelity simulators [Caird and Horrey, 2011]. Figure 2.5 illustrates the different types of simulators. Low-fidelity simulators, like the example in Figure 2.5a, use small screens, have no motion platform, and only basic input devices. Moderate-fidelity or medium-fidelity simulators apply more sophisticated visualizations as shown in Figure 2.5b. For example, large screens or even virtual reality head-mounted displays (VR-HMD). They sometimes have motion cues like vibrating seats or small 3 degree-of-freedom (DOF) motion platforms. High-fidelity simulators often have complex 6- to 9-DOF motion platforms, advanced visualizations like projection-based 360° displays, and spatial audio, illustrated in Figure 2.5c. Another important factor when describing simulators is the visual quality (or graphic fidelity) of the presented visual output. Results of Merenda et al. [2019] indicate that graphic fidelity impacts driving performance and gaze behavior by requiring different patterns of divided attention than simulators with low graphic fidelity. Moller [2011] adds that low graphic fidelity might negatively impact presence—the feeling of being there—in a driving simulation. Similar to the graphic fidelity, the quality of other sensory cues needs to be of appropriate quality to achieve a valid driving simulation. This includes audio and motion, but also for example feedback behavior of steering wheels and pedals.

Overall, the more cues and the better the quality of the cues (audio, visual, and haptic), the higher the fidelity of the driving simulator [Greenberg and Blommer, 2011]. Simulator fidelity is often proportional to financial costs, which requires a careful selection when building a simulator.

In addition to the hardware configuration, a complex suite of software and tools is necessary to provide appropriate scenarios. This suite has to model, among others, vehicle physics, traffic including cars, pedestrians, and bicycles as well as road conditions and weather. It should also offer easy scenario design to ease the test and development process. Commercial tools such as IPG CarMaker[1] or STISim[2] provide excellent vehicle physics and traffic models. However, they are only customizable within the scope of the application programming interface provided by the developers. While open-source tools like OpenDS[3] or Carla[4] provide high degree of customization, they are often outdated or special purpose simulators.

Regardless of hard- and software of the simulator, many people in a driving simulator experience symptoms of simulator sickness. Simulator sickness is a syndrome that describes a group of symptoms associated with the usage of any simulation environment like flight or driving simulators. Kennedy et al. [1993] mention three groups of symptoms: oculomotor, disorientation, and nausea. They manifest themselves in, for example, fatigue, headache, eyestrain, or stomach awareness. The

---

[1] https://ipg-automotive.com/products-services/simulation-software/carmaker/ IPG CarMaker, 2020-09-24

[2] https://stisimdrive.com/ STISim, 2020-09-24

[3] https://opends.dfki.de/ OpenDS, 2020-09-24

[4] https://carla.org/ Carla, 2020-09-24

problem with simulator sickness is that it can negatively influence the validity of simulator experiments: if a participant experiences symptoms to a significant degree, it can negatively impact the ability to control a vehicle or a secondary interface. Simulator sickness is similar to motion sickness, but the latter requires stimulation of the vestibular system whereas simulators do not necessarily do that.

There are various theories that try to explain simulator sickness. Among others:

- Sensory Cue Conflict Theory: The perception system receives cues that are not perfectly synchronized or holistic (e.g. participant sees that they move but the vestibular system does not feel any movement). The perception system cannot resolve these conflicts and causes the symptoms.
- Poison Theory: The human body recognizes the conflicting and non-synchronous cues and treats them as poison. Hence, it reacts in a similar way as it would when expelling an emetic substance.
- Postural Instability Theory: This theory is based on the fact that the human body constantly tries to maintain a stable posture. The cues of a simulation disturb this process which might be a cause for simulator sickness symptoms.
- Rest-Frame Theory: Here, it is argued that the symptoms occur due to the mismatch between the frame of reference and the moving driving simulator (e.g. in a simulator with a small motion-platform but no 360° view).

It is important to note, that none of these theories completely explain simulator sickness, but they provide important hints to inform simulator design regarding field-of-view, screen type and layout, laboratory setting, motion cueing, and participant selection [Stoner et al., 2011]. Other theories exist, a good overview is given by Jerald [2016]

2.4   Manual & Automated Driving

In recent years, driving automation has been rising. In 2014, the SAE introduced the standard J3016—a classification for driving automation systems [SAE International, 2015]. The standard has 6 levels, going from 0, *no driving automation*, to level 5, *full automation*. Table 2.1 illustrates the levels of automation with the classification criteria. Level 0 describes traditional manual driving where all control and supervision is performed by the human. Level 1 describes systems where either the lateral or the longitudinal driving task is executed by the system (e.g. adaptive

Table 2.1 Levels of automation according to the SAE. With levels 1–3, the driver has to always pay attention to handle critical events. With level 3–5, the system monitors the environment and initiates actions in critical situations. The system takes over more and more responsibility in higher levels. Adopted from SAE Standard J3016 [SAE On-Road Automated Vehicle Standards Committee, 2014]

SAE level	Name	Steering/ acceleration & braking	Monitoring	Fallback responsibility	Driving Mode
0	No Automation	human driver	human driver	human driver	n/a
1	Driver Assistance	human driver & system	human driver	human driver	some
2	Partial Automation	system	human driver	human driver	some
3	Conditional Automation	system	system	human driver	some
4	High Automation	system	system	system	some
5	Full Automation	system	system	system	all

cruise control), the human has to monitor the system, and react if the system fails for any reason. Automation systems with level 1 can be activated in some situations (or operational design domains). Vehicles equipped with automation systems of level 2 execute the dynamic driving task completely (longitudinal and lateral), but the human still has to pay attention all the time and react on his or her own, especially in case of a system failure (e.g. adaptive cruise control paired with active lane-keep assist). Similar to level 1 systems, it can be used in some situations. Starting with level 3 automation, the vehicle performs the complete dynamic driving task and takes over the monitoring task. The human does not have to monitor the environment but must stay alert. The driver still has a fallback responsibility. Again, this level can be used only in the domains it has been designed for. With level 4 automation, the fallback responsibility shifts to the system. In case of a critical situation that happens while the vehicle is driving in its operational design domain, it either handles the situation safely or gives the human a comfortable amount of time to take over control. Hence, the driver can completely disengage from the driving task. Level 5 automation does not require any human input or supervision to perform the driving task and works—per definition—everywhere.

To this time, there is no commercially available car that fulfills the promise of level 4 or 5 automation. However, companies like Waymo[5] are testing prototypes of

---

[5] https://waymo.com/ Waymo, 2020-09-23

highly automated vehicles. Vehicles with level 3 automation (conditional automation) are commercially available, e.g. the Audi A8[6]

The driving simulator environment presented in Chapter 4 is capable of various levels of automation. In Chapter 6, we apply a level 2 automated system. In Chapter 7 and Chapter 8, a level 3 automation system is used. In Chapter 9, we simulated a level 4 automated vehicle.

---

[6] https://www.audi-mediacenter.com/en/on-autopilot-into-the-future-the-audi-vision-of-autonomous-driving-9305/the-new-audi-a8-conditional-automated-at-level-3-9307   The new Audi A8—conditional automated at level 3, 2020-09-23

Related Work

3

This chapter outlines the current state of automotive user interfaces and by that, highlights challenges that are still open and some that the automotive UI research community might encounter in the future. This chapter also presents the state of research on in-car displays and S3D outside and inside the automotive domain. By combining these two aspects, we motivate why S3D displays might be able to be part of a solution for upcoming challenges.

3.1   In-car Interaction

The introduction of this thesis briefly touched the evolution of automotive user interfaces. From having only primary input devices like steering wheel and pedals as well as no output devices, vehicles are now at a state where they are equipped with a plethora of devices. These devices allow for the interaction with the manifold functionality of the car: driving itself, advanced driving assistance systems, navigation, and entertainment.

Traditional in-car user interfaces are mostly based on haptic controls like buttons and knobs. Such controls are easy to learn and easy to use but provide limited flexibility. The introduction of touch screens, while successful and flexible, introduced a device that requires relatively high visual attention [Ayoub et al., 2019]. Researchers try to incorporate glance free interaction [Beruscha et al., 2017] or mid-air gestures [Ahmad et al., 2014] to remove eyes-of-the-road time and increase safety. Voice interfaces are often used for warnings and notifications in general [Lewis and Baldwin, 2012]. Further, they are integral part of conversational interfaces and dialogue systems (e.g. Ekandem et al. [2018]). They promise to reduce visual demand and

© The Author(s), under exclusive license to Springer Fachmedien Wiesbaden GmbH, part of Springer Nature 2021
F. Weidner, *S3D Dashboard*,
https://doi.org/10.1007/978-3-658-35147-2_3

distraction by presenting information but also by reacting to commands using machine learning-based natural language understanding. In addition to the microphone(s), the speaker setup and quality needs to be sufficiently good to make the presented audio understandable [Simmons et al., 2017]. Similar to gesture control, the recognition quality is key and can be impaired by external and internal noises [Pruetz et al., 2019]. Some researchers try to incorporate the driver's state (e.g stress) into the interaction process. For example, to detect stress [Munla et al., 2015] but also tiredness [Zhenhai et al., 2017] or anger [Wan et al., 2019] of drivers. Using such information, the developed user interfaces try to, for example, lower the stress level [Çano et al., 2017] or recommend a pause [Aidman et al., 2015]. Another UI modality applied in cars is haptic interfaces. For example, vibration motors in the steering wheel, in touchscreens or in the seat. More advanced, ultra-sonic sound has also been used to provide feedback for mid-air gestures [Long et al., 2014]. Such interfaces can be used to support touch-screen interaction [Beruscha et al., 2017] but also to successfully provide simple alerts in critical situations [Enriquez et al., 2001]. However, classical vibration feedback in various locations can be hard to interpret [Väänänen-Vainio-Mattila et al., 2014]. But it has been suggested that it performs good in combination with other modalities [Kern and Pfleging, 2013].

To reduce workload and distraction, many researchers try to present warnings in a combined fashion: e.g. by pairing visual, auditory, and tactile information in one warning. Such multimodal interfaces have been shown to lower workload and increase performance [Politis et al., 2013; Swette et al., 2013]. Key challenge is to design the interface in such a way that cues are not conflicting and do not lead to sensory overload [Ayoub et al., 2019]. This is because multimodal input has been shown to degrade performance by accidental interaction or mode confusion [Ayoub et al., 2019].

All these devices were predominantly researched and designed for the domain of manual driving. Their general objective is to support the driver during driving. The UIs were designed to minimize driver distraction and maximize safety. While this challenge is still valid in the domain of automated driving, various other challenges have emerged.

3.2   Visual Displays

Over the years of automotive UI development, displays have gotten more and more important. Initially, only few analogue gauges showed speed and fuel level to drivers. Nowadays, digital displays present all kinds of vehicle information but also navigation, weather, air conditioning, entertainment, and many other data. This is due to novel technological possibilities but also because the plethora of functionality in modern vehicles cannot be conveyed by traditional displays.

Especially touchscreen displays gained popularity. They can offer a high resolution and a good refresh rate. Also, they allow for a direct manipulation of control elements. Usually, displays in cars can be divided by their location. Kern and Schmidt [2009] proposed a scheme for this classification . Behind the steering wheel, the instrument cluster resides. Right of the steering wheel is the vertical center console. It can be divided into the upper and lower center console. Also, some cars feature displays at the horizontal center stack. In addition to that, there are some vehicle interiors that have displays left to the steering wheel and on the right side of the passenger area, e.g. for digital rear-view mirrors [Large et al., 2016]. Further, sometimes the area in front of the passenger seat is a display for passenger interaction [Inbar and Tractinsky, 2011]. Researchers and industry also look at other displays, facilitating the windows of the car. (Augmented reality) HUDs [Lisle et al., 2019; von Sawitzky et al., 2019], windshield displays [Ng-Thow-Hing et al., 2013; Wiegand et al., 2018], and augmented passenger or side windows [Häkkilaä et al., 2014; Matsumura and Kirk, 2018] have been investigated to support drivers. Especially Augmented Reality-HUDs have gained attention and are thought of as a viable candidate to extend the user interface of the vehicle [Tönnis et al., 2006; Wiegand et al.,2019]. Another technology that uses various spaces of the car to communicate information is ambient lighting. Various combinations and placement strategies of LEDs have been evaluated to support drivers in lane changes, guide attention, and communicate system behavior [Borojeni et al., 2016; Löcken et al., 2015].

Summarizing research on various visual in-car displays, it is noticeable that the amount of output devices has increased — from few gauges to many displays — and might increase even more, considering the usage of head-up and augmented reality displays. In this thesis, we take up this trend and combine the areas left of the steering wheel, the instrument cluster, and the vertical center console into a single, seamless, and large display. This display provides an interactive surface — similar to a tabletop. Such a device could, for example, adapt to different driving modes and activities but also present information at the very location the driver (or passenger) is looking at. Also, it could increase the design space and allow for novel dashboard designs and applications. With the increased design space, it could provide possibilities to tackle the mentioned challenges.

3.3   S3D User Interfaces

3.3.1   S3D Across Domains

Read [2015] state that, in nature, stereo vision provides an evolutionary advantage. It helps to detect anomalies in cluttered scenes, like leafy backgrounds, and while judging relative depths. Also, stereo vision supports execution of visually guided manual tasks, e.g. using tools. It is, however, a rather poor help when judging absolute depths.

Research and industry alike have applied and evaluated stereoscopic 3D across many domains. While the previous chapter outlined various monocular user interface technologies that have been researched to improve the in-car experience, the following paragraphs focus on S3D. We first cover general application areas that benefit from S3D and then progress towards specific previous use cases in the automotive domain.

For virtual scenes, Ragan et al. [2013] confirmed that S3D can help performing small scale distance judgment tasks. In their experiment, participants had to navigate through a 3D maze and count certain items. In addition to S3D, they added head-tracking and found that the combination of both lead to the best performance in the task (1280 × 1024 px, CRT projector; active 3D; 10' × 9' display). In another maze-like task, participants were asked to solve a maze in a multi-level floor plan that was either in S3D or in 2.5D. With S3D, they were not only faster but also preferred the added depth cues (8" tablet, autostereoscopic) [Rantakari et al., 2017]. Similar, De Paolis and Mongelli [2014] showed that S3D can increase performance when finding targets on maps and judging distances between them. Participants made fewer errors, were faster, and reported higher presence, better depth impression, and increased situation awareness when S3D was enabled. However, the viewing comfort suffered with S3D (1920 × 1200 px, 12 m^2, active 3D projection). Boustila et al. [2017] showed that this also translates to architectural visualizations. While observing architectural scenes participants had to estimate distances, sizes, and rate the habitability of the scene. Size perception and distance estimation as well as the judgment of habitability were better with S3D. A monoscopic visualization without head tracking performed worst. Adding head-tracking to a monoscopic visualization still improved results. However, they were still worse than the S3D visualizations with or without head tracking (active 3D projection, 120 Hz, 1400 × 1050 px, 3 m × 2.25 m). In all these examples, S3D enhanced the understanding of spatial spaces.

Not only in the architectural domain, but also in CAD-related activities — which are inherently spatial — S3D has been shown to provide benefits in depth estimation, judgment of depth differences, and in mental rotation of objects (3 m × 3 m, active

3D, 960 × 960 px) [Gaggioli and Breining, 2001]. It is important to note that the authors also investigated model complexity (low vs. high) and mode of model presentation (e.g. wireframe vs. full shaded). For both, S3D outperformed 2.5D.

Next to architectural and general understanding of map-like data, medicine and therapy have been a rich field for S3D applications. For example, while the overall speed of image-guided laparoscopies supported with S3D has not been increased in general, performance increased with the difficulty of the procedure (1920 × 1080 px, passive 3D) [Schwab et al., 2019]. Similar to that, S3D lead to a better detection rate of aneurysms in computer tomography scans and promises to be a valid alternative to traditional methods (1920 × 1080 px, 46", 60 Hz, passive 3D) [Acker et al., 2018]. S3D is not only useful in diagnostic and invasive procedures, but also for therapy that is based on videos like the treatment of amblyopia (no display specs provided) [Herbison et al., 2015].

S3D has also been applied in the area of video conferencing. Researchers investigated the influence of S3D when applied to the video stream. They found that images were perceived more realistic, and that presence was rated higher (720 × 480 px, 120 Hz, active 3D, 21" screen) [Tam et al., 2003]. Hemami et al. [2012] did follow-up research and found that, while S3D leads to better depth perception, it does not necessarily lead to a better ability of the partners to communicate or a better quality of the two-way interaction. They did also not measure an increase in perceived usefulness (life-sized display, passive 3D).

Another domain where S3D has been heavily researched is the gaming area in specific and entertainment in general. [2012] let people play three different games (*Avatar: The Game*, *Blur*, and *Trine*; 1680 × 1050 px, 120 Hz, active stereo). Their results indicate that presence is enhanced but so is simulator sickness. It is noteworthy that, while simulator sickness increased for all three games, presence was only higher in *Avatar: The Game* and *Blur* — both are games where depth cues are more important whereas *Trine* is a 2.5D side-scroller that does not rely much on depth cues. However, participants' qualitative feedback indicates that in all three games, S3D seemed to lead to a more thoughtful and natural experience. The importance of depth cues for a manifestation of S3D's benefits have also been shown by Kulshreshth et al. [2012] (active stereo Asus ROG Swift 3D, 27", 2560 × 1440 px, 144 Hz). Here, the authors claim that isolated interaction with few objects and a static background are key requirements for S3D to show its benefit. Recent results of Hossain and Kulshreshth [2019] support these findings. In their experiment, S3D led to higher engagement and excitement but also presence. Their participants had to play five games on an S3D monitor (2560 × 1440 px, 144 Hz, active 3D) with S3D activated and disabled. Again, in games where depth was part of the gameplay (*Left4Dead* and *Portal 2*), players achieved higher scores than in the 2.5D condition.

Besides gaming, S3D has been shown to be beneficial for other types of entertainment like watching television. Yang et al. [2012] investigated how S3D influences watching a movie on a 3D television (55", 240 Hz, 1920 × 1080 px, active 3D). Here, S3D provided greater immersion but also lead to more visual discomfort and motion sickness. Their results indicate that the effects depend on viewing angle and distance: the higher the viewing angle, the more symptoms, and the less immersion.

Dixon et al. [2009] analyzed the potential of S3D in the military domain. They mention that S3D can lead to more stress, visual fatigue, and simulator sickness. However, they also note that S3D can increase presence, lead to faster navigation, and increased situation awareness. Finally, they also take into account the costs that were necessary at that time (2009) to realize stereoscopic 3D. Considering this, they concluded that the benefits do not outweigh the cost. It is important to note that recent advances in 3D display technology and the accompanying reduced cost might lead to a different outcome.

In the automotive area, S3D has also previously been applied outside the domain of user interfaces. S3D can improve presence in driving simulations when applied to the environment simulation without impairing vection, involvement, or simulator sickness [Ijsselsteijn et al., 2001] (50° hFOV, 1.90 m × 1.45 m). Also, Chen et al. [2010] investigated the influence of S3D on driving performance when operating a remote vehicle (teleoperation). It was found that active stereo led to faster completion of the route compared to passive stereo. Again, there was no significant difference in simulator sickness or workload compared to 2.5D (22", 120 Hz, 1920 × 1080 px, active and passive 3D). Forster et al. [2015] investigated a driving simulation with enabled and disabled stereoscopic 3D (220° hFOV, 6.3 m × 2.8 m; 120 Hz passive 3D). They found that presence is increased in an S3D driving simulation. However, participants also reported more simulator sickness. Regarding distance estimation, S3D played only a minor role in their setup. There were no improvements of verbal distance estimation, but only for distance adjustment while driving.

Overall, S3D has been shown to have several benefits across many domains — including driving. However, it also leads to increased symptoms of simulator sickness or visual fatigue. In 2014, McIntire et al. [2012] performed a comprehensive analysis of S3D displays that focused especially on human-computer interaction. This survey was extended in 2015 to other domains [McIntire et al., 2014; McIntire and Liggett, 2015]. In their analysis, they categorized studies not by domain but by the activity the task represented. Figure 3.1 shows the categories, the number of studies, and the aggregated results. For all categories except "learning, training, planning" and "navigation", S3D scored better than 2.5D in more than 52% of the studies. Hence, the authors conclude that S3D is especially beneficial for tasks that include the "judgment of positions and/or distances", "finding, identifying or

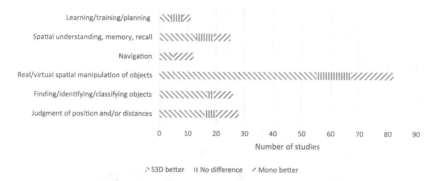

Figure 3.1 Overview of the benefits of S3D on various tasks. Adopted from [McIntire et al. 2014]

classifying objects", and "real/virtual spatial manipulation of objects". Also tasks that require "spatial understanding, memory, and recall" seem to benefit from S3D. However, they also note that S3D is less beneficial (or not at all) if the task has been learned already and for those tasks that do not inherently require depth cues for successful completion or good performance. Related, tasks and applications that already have strong depth cues — aside form S3D — or where the important binocular depth cues are outside the effective range (c.f. Figure 2.2), might not benefit much from S3D. In return, they report that S3D helps most if the depth-related task is in the near-field (and within the effective region of binocular depth cues). By analyzing the experimental tasks, they also concluded that S3D helps with difficult and complex tasks like visual searches and precise 3D manipulations, meaning that it helps with spatial or multi-dimensional tasks where manual interaction with objects is necessary. S3D also seems to support the understanding of complex or even ambiguous scenes (similar to the mentioned function of binocular vision in nature to break camouflage) and in general for data and information visualization. Also, they conclude that S3D is especially helpful for tasks that can be classified as difficult, complex, unfamiliar, or where monocular depth cues are degraded.

Overall, S3D seems to be beneficial in several application and task domains from medical to driving simulation. Quite often, the drawback of increased visual fatigue is mentioned — the former is a strong motivation for our research whereas the latter reminds us to pay attention to simulator sickness symptoms. After having established a high-level overview of benefits and disadvantages of S3D in various domains, the following chapter will outline the previous research on S3D displays in cars and the implications that arise from this research.

Figure 3.2 3D rendering an in-car S3D display by Continental (© Continental)

3.3.2 Automotive S3D UIs

Figure 3.2 illustrates the current vision of Continental and their large *natural display*. While companies like [Continental AG 2019, Robert Bosch GmbH 2019], or Peugeot Peugeot Deutschland GmbH [2019] currently advertise S3D displays with better comprehension of information, better design, and lower reaction times, research on automotive S3D displays is scarce.

In 2008, Krüger[2008] evaluated if binocular depth cues can help maintaining distance to a leading vehicle. In a medium-fidelity driving simulator experiment with an ASD display (640 × 600 px) as an instrument cluster, participants had to estimate distances on the ASD while driving a vehicle with an adaptive cruise control system. Interestingly, distance estimation was worse with the ASD and there was no clear favorite when comparing results to 2.5D visualizations. The ASD also lead to higher variation in steering angle which could indicate problems in keeping the car on the road (in their experiment, this depended also on the style and viewing angle). Also, participants did not rate the ASD as being more attractive or useful. However, the ASD attracted more short glances, suggesting that participants spent less time looking on the instrument cluster compared to the monoscopic version.

Later, Broy et al. [2012] used an autostereoscopic laptop (Asus G73JW 3D, 1280 × 480 px, 120 Hz, active 3D) in a low-fidelity driving simulator study without driving task. The laptop represented the center console and participants had to operate a music player, a navigational system, and change settings while keeping an eye on a target displayed on the front screen. Participants reported increased hedonic quality and attractiveness with S3D, and that information was easier to perceive. However, a clear benefit of S3D compared to 2.5D regarding peripheral detection performance or workload could not be shown. In another experiment, Broy et al. [2013] investigated the design of such displays further. In a low-fidelity driving simulator, a laptop with S3D display (Asus G73JW 3D, 1280 × 480 px, 120 Hz, active 3D) replaced the instrument cluster. By adjusting squares on the S3D display, participants set the maximum and minimum disparity. In the end, the authors defined a zone of comfort of 35.0 arc-min (or 0.6°) of angular disparity. They also report a minimum depth difference between two objects (or layers) of 2.7 arc-min (0.045°). However, they note that the zone of comfort is highly individual. In the conclusion, they also add that designers should keep a reference object at screen level with a disparity D = 0°.

A further experiment by Broy et al. [2014b], executed using a low-fidelity driving simulator with an S3D instrument cluster (Asus G73JW 3D, 15.6", 1280 × 480 px, 120 Hz, active 3D), investigated the influence of motion parallax via head-tracking on aesthetics and readability of various designs. Quantitative results indicate that S3D without motion parallax increases UX, readability, and usefulness but S3D with motion parallax does not. Also, they concluded that the depth cues often influence perceived quality of the display more than the actual UI designs. Considering the qualitative feedback of the participants, the authors add that motion parallax provided intuitive zoom to participants but also that S3D in general could be a dangerous distraction. Regarding the negative quantitative performance of motion parallax, the authors close the article with the outlook that it should be investigated further due to poor performance of the hardware.

In another medium-fidelity driving simulator study by Broy et al. [2014a], the instrument cluster was replaced by two different notebooks to test various 3D technologies (active 3D: Asus G73JW 3D, 15.6", 1280 × 480 px, 120 Hz; passive 3D: 15.6", 1366 × 768 px, 120 Hz). They reported increased task performance while judging depths (e.g. to a leading car or with navigational cues) and while getting urgent content on an auto-stereoscopic display. Another finding was that S3D works best with low complexity elements and high-quality displays.

Szczerba and Hersberger [2014] reported results of a desktop study within a driving context (no driving task). The authors placed their participants in front of an auto-stereoscopic screen (1920 × 1200 px, 120 Hz) which showed an instrument cluster. Results indicate that eyes-off-the-road time depends on disparity — negative disparity (objects in front of the display) lead to less eyes-off-the-road time. Also, participants were more accurate when identifying displacements of objects on an S3D display. They also report that perceived image quality degraded with increasing disparity and that visual discomfort increased when using S3D. Results indicate an increased search performance, but only when the target object was absent.

A real-world driving study with a car that had an S3D display mounted behind the steering wheel to provide an S3D instrument cluster revealed several findings regarding workload and user experience (UX; 13.3", 1920 × 1080 px, ASD). Interestingly, there was no influence of S3D on workload compared to a monoscopic display. However, participants reported a better information structure and user experience, but also increased simulator sickness symptoms. They also reported higher perceived attractiveness [Broy et al., 2015a].

In another real-world driving study, domain experts (concept designers of automotive UIs) had to react to notifications while driving and were observed by the experimenters [Broy et al., 2015b]. The instrument cluster was a S3D display (13.3" ASD, native resolution: 1920 x 1080 px; 3D resolution: 1114 x 626 px for each eye). Qualitative feedback revealed the positive effects of S3D on acceptance, aesthetics, and usefulness but also the possible disadvantage of visual discomfort and potential problems with the technology in the car, e.g. tracking and reflections. Readability tended to be judged worse in S3D compared to 2.5D. Especially the better information structure (more important content is closer) was welcomed by participants. But, it was also sometimes not clear which UI elements belong together (an aspect of less than optimal design as the authors acknowledge). The experts mentioned that depth could especially be good to visualize spatial and temporal relationships.

Pitts et al. [2015] report results from a high-fidelity driving simulator study with three experiments (10.1", ASD). In the first study, participants had to detect an object in a group of three that had a different disparity than the other two (change detection). S3D lead to a faster completion time, which manifested itself in lower eyes-off-the-road time. In the second study, they wanted to investigate the influence of disparity on perceived image quality. Results indicate that image quality degrades the more disparity deviates from $0°$. This is also true for disparity values within the zone of comfort. With lower perceived image quality, participants reported increased visual discomfort. Motivated by that, the authors investigated what influences the minimum disparity that leads to a perceptible 3D effect. They conclude that this depends on the size of objects, their position in space, and proximity to other objects which act as a reference and by that, confirm statements of Howard and Rogers [2008].

Dettmann and Bullinger [2019] investigated if ASDs can increase driver's performance when judging traffic situations. In a low-fidelity simulation, they showed participants traffic situations on an ASD (13.3 in, 1920 x 1080 px) and on a traditional monoscopic display. Results indicate that S3D can lead to a more precise judgment of position. Contrary to other results, there were no measurable effects on visual fatigue. However, participants were not required to drive but were only presented videos which might limit validity of the results.

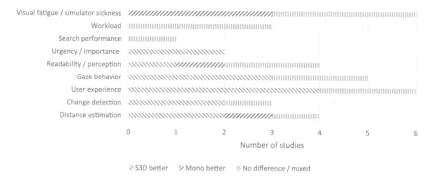

Figure 3.3 Overview of S3Ds' benefits in the automotive domain on various tasks (N=11, some studies investigated two or more aspects)

In another study, Dettmann and Bullinger [2020] evaluated the application of an ASD (13.3", 1920 × 1080 px) as output for a rear-view camera during reverse parking. Participants had to perform a reverse parking maneuver and rely on a rear-view camera that was either an ASD or a traditional monoscopic display. Neither performance, nor user experience or workload were affected by the display technology. The absence of benefits is, according to the authors, due to the low resolution of the stereo camera but also due to the low task complexity — the task was performed in a closed environment without any risks of damage or collision.

Figure 3.3 summarizes the results of the previous work on automotive S3D displays. Note that the displays in these studies had often a small form factor and replaced usually either the instrument cluster or the center console. The summary confirms that visual fatigue and simulator sickness are major hurdles of S3D displays — probably due to sub-par hardware specifications like resolution or refresh rate. However, it is not inherent to S3D displays as 50% of the studies have not led to increased symptoms. Besides that and except for *workload* and *search performance*, S3D either showed some benefits over 2.5D displays. Especially the *communication of urgency or importance*, *gaze behavior* (e.g. eyes-off-the-road time and mean glance duration at visual stimuli), *UX*, and the *change detection performance* has been shown to be positively affected by S3D.

3.4   Challenges of in-car User Interfaces

In recent research, Ayoub et al. [2019], Boll et al. [2019], Kun 2018], and Riener
et al. [2016] outline core challenges arising with the advent of automotive driving. Among others, improved usability, a design process with heterogeneous user
base, real-time trust estimation, passenger interaction, accessibility of vehicles for
marginalized groups, legal issues of automated driving, granularity of control, or
remote operation. For driver-related tasks, these reviews describe trust, take-over
processes, and NDRTs (non-driving related tasks) as especially important research
topics. In the following, we will outline these topics further.

Trust is important for automated driving. With increasing level of automation,
the car takes over more and more of the driving task. By that, the human hands
over more and more control to the automation system. It is inherently necessary
for this cooperation, that the human has sufficient trust in the vehicle and accepts
that it is in control [Endsley, 2017]. If the level of trust is not high enough, it is
possible that the automation system is not used at all or — maybe worse — used in
an ill fashion Lee and See [2004]. To establish trust, [Lee and See, 2004] suggest to
continuously monitor drivers to get information about what they are doing. Using
this information, the system should offer transparent information on what the car is
doing. They also suggest that the automation system should be an emotional system
that reacts to the driver's state.

The issue of take-over maneuvers is a highly important one because it can lead to
potentially fatal situations. In theory, the human has to supervise a level 3 automated
vehicle and only needs to react if the car issues an alert. However, it is likely
that humans — now passengers — are out-of-the-loop, meaning that they do not
fully commit to supervising, that concentration is low, and that distraction is high
[Endsley, 2017]. Automotive user interfaces can support people to correctly react
in such situations and by that, avoid potentially fatal situations. Various mono- and
multimodal techniques incorporating auditory [Politis et al., 2015], visual [Melcher
et al., 2015], and haptic cues Borojeni et al., 2017] have been researched to speed
up the take-over process and to enhance performance. However, the problem is
complex due to almost infinite situations in which a take-over can happen.

Another challenge that arises with advancements in automated driving is that
driver-passengers can potentially more and more reclaim the time spent driving
and use it for other non-driving related activities. Especially work- and well-being
related tasks could enhance the experience [Ayoub et al., 2019; Boll et al., 2019; Kun,
2018]. However, vehicles are not necessarily designed and suited for such NDRTs
[Large et al., 2018]. In a study, Oliveira et al. [2018] showed that participants mainly
preferred in-vehicle displays for NDRTs over their own smartphone. Schartmüller

et al. [2018] showed how modifications to the steering wheel might be necessary to support typing activities. These studies highlight, that users want to perform NDRTs, and that they might prefer in-vehicle displays, and that the current input-devices (like a steering wheel), are not necessarily appropriate for such tasks.

As the related research on automotive user interfaces shows — despite the almost 150-year long history — manual driving still poses many problems to research and industry. In addition to that, new challenges have been arising with automated driving. Many researchers point out that, among others, trust-evoking UIs, approaches to NDRTs, and UIs that handle control transfer are necessary to continue the success of automated vehicles and to make driving (or being a driver-passenger) a joyful but also efficient and safe experience. Hence, these are the use cases where this thesis exemplary applies S3D dashboards to explore if and how automotive UIs can benefit from binocular depth cues in these areas.

3.5   Implications & Summary

The review of the related work shows that over the years S3D displays have been explored in many applications, scenarios, and tasks. In many non-automotive domains, it led to promising results across many task categories. According to McIntire et al. (c.f. Figure 3.1), especially the judgment and understanding of spatial information, visual object processing, and the manipulation of spatial situations seem to profit from binocular depth cues. In addition to that, there is a large body of work that shows that S3D can increase presence during forms of visual communication like video calling. In the vehicle interior, S3D displays have been explored as replacement for the instrument cluster the center console — overall with rather promising results.

Data from all these domains and the positive results in vehicles motivate a further exploration in the automotive domain. Previous work shows that S3D could enhance UX in general and the communication of urgent or important information on the small displays. However, it is unclear how it performs in the peripheral field of view and on other dashboard locations where it has the potential to replace traditional displays. With driving being an inherently visual-spatial task, the added depth cues of S3D might be able to support this activity by providing, for example, navigational information. The benefit of S3D in the manipulation of real/virtual objects might also be an aspect that can be facilitated in vehicles for NDRTs.

Hence, this thesis describes the exploration of large S3D dashboards for future challenges in the automotive domain. Based on the information presented in Section 3.4, it will focus on the driver-centered challenges of automotive UIs:

(1) — Reduce driver-distraction
(2) — Increase Trust in Automation
(3) — Increase performance during take-over
(4) — Enable execution of NDRTs

We will target the usage of S3D to support in-car interaction (1), as a way to use
S3D to enhance trust (2), to present navigational information during take-over (3),
and to explore the interaction with S3D content in cars (4). We do this by means of
a large dashboard that covers the reachable area of a driver starting from the left of
the steering wheel towards the right side of the vertical center console. To do this,
a driving simulator will be designed that allows for the application and evaluation
of large interactive S3D dashboards.

Part II

Design & Development of S3D Dashboards

Prototyping S3D Automotive UIs

4

> Parts of this section are based on Weidner and Broll (2017a), Haupt et al. (2018), and Weidner et al. (2019).

To develop and evaluate stereoscopic 3D user interfaces for cars, a setup is necessary that helps to make informed judgments about this technology. Aim of this section is to outline the development process and the final design of the integrated driving simulator environment used in this thesis. First, a short overview about prototyping setups for automotive UIs is given. After that, we outline the soft- and hardware components as well as the development process of the simulator environment used in this thesis. Following that, a brief discussion of our decisions follows.

4.1 Problem Statement

First and foremost aspect of this research is to explore if a large interactive 3D display provides benefits for drivers. Given this core objective, the prototyping setup needs to fulfill several requirements. To maximize research possibilities, the prototyping setup must be able to provide binocular depth cues on a large display — at best covering the available interaction space of the driver. The display providing these cues should make use of state-of-the-art hardware when possible and avoid subpar devices that offer outdated screen resolutions and refresh rates to minimize simulator sickness. The setup should also be able to allow for the exploration of the scenarios mentioned in the previous chapter (c.f. 3.5, driver distraction, trust, take-over, and non-driving related task). To explore interactivity, the design should not prevent the integration of various interaction techniques in this space (e.g. via hardware that obstructs movement or vision). The experimental nature of large and interactive S3D dashboard is likely to make rapid changes and redesign of the hard- and software environment necessary. Hence, the development environment must be able to evolve with changing requirements. It should also allow for the extension with new input and output devices. Further, it should make it easy to deploy new non-functional designs to assess early prototypes. Next to an S3D screen and interaction modalities, the whole system still needs to offer driving functionalities and should be ready to operate with various levels of driving automation. In this context, it is important that the UI software receives data from the driving simulator, hence a connection between the S3D dashboard and the driving simulator software is necessary. Finally, because user studies are fundamental for the exploration of this in-vehicle user interface, the environment should be usable by a wide range of participants (e.g. people with various body sizes) or should at least be able to adapt to a wide range of participants.

Overall, the core requirements of the prototyping setup can be summarized as follows:

- Large S3D dashboard: a display that provides binocular depth cues to the driver on a large area.
- Graphic fidelity: hard- and software should provide state-of-the-art graphic fidelity.
- Interactivity: it should be possible to interact with 3D content by means of various input modalities.
- Reconfiguration: addition and removal of input devices but also possible changes on the driving simulator should be and supported.
- Connectivity: the S3D dashboard should be able to interconnect with a driving simulation to process and exchange relevant data.
- Rapid prototyping: test and development of new UI prototypes in various application and driving scenarios should be possible and require only short development cycles.

4.2   On Automotive UI Prototyping Environments

In the following, we briefly review recent prominent and existing tools that have been used for the design and development of AR/VR or S3D in-car user interfaces and evaluate them with respect to the above-mentioned categories.

Gonzalez-Zuniga et al. [2015] presented a tool based on a web browser that allows to create UI mock-ups in S3D. However, their solution does not support high-fidelity graphics or the connection with any interaction devices or driving simulator. *Skyline*, the prototyping setup presented by Alvarez et al. [2015] relies on a component-based approach. Its hard- and software is designed for rapid prototyping which makes it ready for the integration of in- and output devices. However, it does not support large S3D due to their specific hardware setup. Also, it is not open source and was not available for purchase at the time of development. Chistyakov and Carrabina [2015] present a JavaScript library that allows for the creation of S3D mock-ups called *S3Djs*. The lightweight nature of JavaScript makes it possible to integrate it into other applications. With additional effort regarding implementation, it could be used as a rapid prototyping application for automotive S3D user interfaces. Similarly, Gonzalez-Zuniga and Carrabina [2015] present *S3Doodle*. It is a browser-based tool for rapid prototyping of S3D user interfaces via hand and finger gestures. While the web-based approach makes the tool versatile, it does not natively offer any integration to a driving simulator or any other input devices. In this regard, it is similar to the initially mentioned tools of Gonz et al. and Zuniga-Gonzalez et al. (no integration of high-fidelity graphics or any readily available interconnection with input or output devices). *OpenDS* offers this connection and also an extension called *OpenHMI* [OpenDS, 2016] that allows for the development of new automotive user interfaces. It has been designed to develop new instrument cluster concepts. However, the overall system provides low graphic fidelity and only limited interconnection to input and output devices. With *autoUI-ML* [Deru and Neßelrath, 2015], Deru et al. proposed a work-in-progress version of a basic markup language for the development of automotive UIs. The authors provided a system incorporating upcoming multimodal interaction techniques such as speech or gesture input, but also include practices based on knobs and 3D controllers. While the concept of a markup language is appealing, the solution does not support binocular depth cues or the design of them and was only considered as a proof of concept without any reference implementation. Important aspect of the mentioned toolkits is that many of them are not publicly available. All except OpenHMI and S3Djs were unavailable at the time. However, because of the low graphic fidelity (OpenHMI) and the bare nature of S3Djs, both were not considered as adequate candidates for our prototyping setup. *Skyline* and its component-based approach informed our design.

Besides tools dedicated to S3D and automotive UIs, many general-purpose tools for the development of interactive systems are available. Two very prominent ones are the markup languages *Virtual Reality Mark-Up Language* (VRML, [Carey and Bell, 1997] and its successor *Extensible 3D* (X3D, [Daly and Brutzman, 2007]). They are markup languages used for describing and developing interactive 3D content and especially tailored towards VR/AR applications. These scene description languages are based on a scene-graph and allow for an easy organization and reor-

Table 4.1 Summary of various prototyping setups for S3D applications and/or automotive AR/VR-based user interfaces("+": supported, "–": not/badly supported.)

	Renders S3D content	Native high graphic fidelity	Supports developing interactive applications	Supports reconfiguration and/or extension		Supports rapid prototyping		Connects to DS (or offers interface)	Open source and/or publicly available
				SW	SHW	SSW	HW		
Gonzalez-Zuniga et al. [2015]	+	–	–	–	–	+	–	+	–
Alvarez et al. [2015]	–	+	+	+	+	+	+	+	–
Chistyakov and Carrabina [2015]	+	–	–	+	–	–	–	–	+
Gonzalez-Zuniga and Carrabina [2015]	+	–	–	–	–	–	–	–	–
OpenDS [2016]	–	–	+	+	–	+	–	+	+
Deru and Neßelrath [2015]	–	–	+	+	+	+	+	+	–
X3D Daly and Brutzman [2007] & VRML Carey and Bell [1997]	+	+	+	+	+	–	–	+	+
Unity[1]	+	+	+	+	+	+	+	+	+
Unreal Engine[2]	+	+	+	+	+	+	+	+	+
Riegler et al. [2019]	+	+	+	+	+	+	+	+	–
Conrad et al. [2019]	+	+	+	+	+	+	+	+	–
Gerber et al. [2019]	+	+	–	+	–	+	+	+	–

ganization of content. While they are quite suitable for visualization of data, they have only a limited set of interaction capabilities, provide rather simple animation schemes [Figueroa at al., 2005], and are constrained regarding extensibility [Olmedo et al., 2015]. With *Unity*[1] and *Unreal Engine 4 (UE4)*[2], game engines have gained popularity for the development of interactive non-gaming applications. While they both provide a similar set of functionality and operate on the same basic principles, Unreal Engine is slightly more open — it allows free access and modification of the source code, making it highly flexible but also feature-rich. These game engines come with editors that allow for the rapid creation of prototypes and the support of various input and output hardware. Unreal Engine has an integrated, although basic stereoscopic 3D mode. Other, very recent, prototyping setups of Riegler et al. [2019], Conrad et al. [2019], and Gerber et al. [2019], are also based on these engines — however, they were developed when this thesis was in its final stages not available at the time development started.

Regarding S3D displays in vehicles for the design of automotive stereoscopic 3D user interfaces (c.f. Section 3.3.2), previous researchers used either rather small ASDs or active/passive S3D systems and replaced one of the traditional displays — either the horizontal center console or the instrument cluster (c.f. Section 3.3.2). Those with available information were developed with Unity or C++ for Android.

Table 4.1 summarizes prototyping setups for automotive user interfaces according to the previously mentioned criteria.

Considering the overall state of prototyping tools, their availability, maturity, and the features they offer, as well as the absence of an available environment that fulfills the derived requirements, this thesis presents the design and development process of an integrated prototyping setup for large S3D dashboards based on game engines, namely Unreal Engine 4. This decision has been subsequently confirmed by the introduction of three other prototyping toolkits based on game engines.

4.3   The Environment Simulation

Figure 4.1 shows the final prototyping setup. It consists of an environment simulation and a car mock-up that features the S3D dashboard. Having a large screen that provides good graphic fidelity, a surround sound system for vehicle and environment sounds, and a car mock-up but no motion platform, the driving simulation environment as a whole can be classified as a medium-fidelity simulator. The fol-

---

[1] http://unity.com/, Unity, 2020-09-23
[2] https://www.unrealengine.com/, Unreal Engine, 2020-09-23

lowing sections explain the details of the hardware and software, focusing on the
buck with the large stereoscopic 3D dashboard.

Figure 4.1 Integrated driving simulation environment showing the front screen and the buck
with rear-view mirror. Here, the large S3D dashboard shows a simple layout with automation
indicator and gauges for speed and rpm

The environment simulation consists of two basic elements. The front screen
showing the actual environment, and the sound system displaying driving and environment cues. The front screen, a Vision3D foil[3] has a size of 3.6 m × 2.25 m. It is
illuminated by a Barco F50 projector (Digital Light Processing, WQXGA — 2560
× 1600 px, 120 Hz) equipped with an EN56 lens (1,14 — 1,74:1). The simulation
environment is powered by a workstation equipped with a Dual Intel Xeon E5-2670
v3, 64 GB RAM, a Nvidia Quadro M6000, 12 GB, and runs Windows 10. The
rear-view of the setup was realized by a BenQ MH740 projector (1920 × 1080 px,
60 Hz, 4000 ANSI Lumen, 2.5D, 1.50:1) and a real car rear-view mirror. This way,
motion parallax was available for the rear-view. The mirror was mounted on a tripod
and could be adjusted by participants to suit their size.

The driving simulation software includes and manages other actors (general
vehicle traffic, environment actors like birds, or pedestrians), roads, and other world
objects like buildings, but also elements necessary for experiments like event trig-
gers and timers as well as all functionality that is necessary for the ego vehicle. The
ego vehicle (or car) is driven by the participant. Here, we extended UE4 with several plugins to fit our needs. In detail, a logging component that produces log files
containing relevant information about actors and events and that can easily be integrated in new driving scenarios using C++ or blueprints. Further, a communication
component that handles sending and receiving data between the driving simulation

---

[3] http://www.3dims.de/001_PDF/vision3D_wideview.pdf, Vision3D foil, 2019-11-18

software and the user interface software (and other software, if necessary). This is realized via TCP/IP. Data is sent using the open format *Google Protocol Buffers*[4]. It allows a key-value-based way of transmitting data, similar to XML, but automatically compresses data to reduce package size. Yet another extension handles the input from the primary input devices and makes them available in UE4. The ego vehicle is also encapsulated in a plugin and by that, can be used in different, new UE4 applications. It contains the behavior of the ego vehicle including the configuration of the vehicle physics (acceleration, deceleration, gearing, suspension, tire stiffness, and slip), and most importantly, the configurations for the various modes of automation. The vehicle is able to perform with no automation, SAE level 2, SAE level 3 and SAE level 4 automation. It is important to note that the implementation is not based on artificial intelligence but rule-based and works within the limits of the respective experimental scenario. While this is easy to use in simple scenarios, complexity of a rule-based system can make it inappropriate for larger scenarios containing many actors and sophisticated behavior. An advantage of the rule-based approach is that it can be set up and extended rather easily without the need for training data and training phases. Future implementations could rely on tools like *Carla*[5] that are designed to bring automated driving to UE4.

The driving simulation application can be run in server mode to support several displays. The server then connects to clients and keeps them in sync. In this case, the server performs all calculations, is responsible for displaying information on the front screen, and for calculating the position of actors and the environment. The client application, in our case the application showing the rear-view, does only receive updates but does not perform any of the major calculations like physics of the ego vehicle or driving automation. While we use one client for the rear-view mirror, future implementations could add several more to achieve a wider field of view or side-view mirrors.

4.4 The Buck

Figure 4.2 shows the initial CAD model of the car mock-up with a 173 cm sized human reference model. The side panel on the right is not displayed to show the design of the motor compartment's interior. We will first outline the single components of the buck and then discuss features, advantages, and open challenges of the mock-up.

---

[4] https://developers.google.com/protocol-buffers/, Google Protocol Buffers, 2019-11-19
[5] http://www.carla.org, Carla Simulator, 2020-09-23

(a) Side view of the buck   (b) Perspective view of the buck

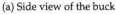

Figure 4.2 Side and perspective view of the bucks CAD model with a 173 cm human reference model

The car mock-up (or buck) is designed to resemble the interior of a 2004 Mercedes A class [M@RS – The Digital Archives of Mercedes-Benz, 2019]. In detail, the position of the steering wheel and the total width and height of the dashboard are similar. The buck is equipped with a gaming steering wheel with force feedback (Thrustmaster TX Leather Edition[6]), pedals, and a 7-gear gearshift (six forward and one rear gear)[7]. While the force feedback is not able to accurately reproduce the characteristics of a real steering wheel, it provides a simplified version of it. It can be controlled from within the driving simulation. The gear shift has no feedback. The pedals provide simple feedback via integrated metal springs. To provide a plausible seating experience, a real car seat from a Volkswagen T5 was installed. The positions of seat, steering wheel, and gear shift can be adjusted along three degrees of freedom (X, Y, and Z). Also, seat and steering wheel can be adjusted in pitch. The pedals are fixed by adhesive tape and can be relocated along the longitudinal axis. By that, the buck allows for adjustment to make the simulator usable for a wide range of participants while initially providing a design inspired by a real car.

Realizing the binocular depth cues on the dashboard would have been possible with various technologies. In general, projection-based, display-based, and HMD-based solutions. In our case, we decided to use a projection-based solution. A display-based solution was considered but was, due to unusual size and form of the desired display, not considered as a viable candidate. While we could have used

---

[6] https://shop.thrustmaster.com/en_gb/products-steering-wheels/tx-racing-wheel-leather-edition.html, TX Racing Wheel Leather Edition, 2020-09-23

[7] https://shop.thrustmaster.com/fr_fr/th8a-and-t3pa-race-gear.html, TH8A & T3PA race gear, 2020-09-23

several smaller displays next to each other, this would have introduced boundaries of a few millimeters between them and would have also required synchronization. For the HMD-based solution, no AR-HMD providing sufficient quality and field-of-view was available. The same was true for solutions based on VR-HMDs. However, recent advances in HMD technologies regarding field of view, resolution, and refresh rate might lead to different design decisions. For the projector-based solution, two approaches were considered: projection on a surface from the top (front-projection) and rear-projection. A projector augmenting the surface from above was discarded. This approach would have probably limited the mobility of the mock-up due to a ceiling-mounted projector. More importantly, it would have led to shadowed areas if an object (e.g. the hand) is between the projector and the projection surface. In the end, we decided to apply a rear-projection to realize the binocular depth cues. The projector is located in the "motor compartment" of the buck. The equipped Barco F50 projector (Digital Light Processing — DLP, WQXGA — 2560 x 1600 px, 120 Hz) is paired with a EN56 lens (1,14 — 1,74:1) and mounted with a tilt of 40°. It does not directly project its image on the projection surface but first on a mirror. The mirror is a first-surface mirror to avoid double images due to refraction. The redirected image is then projected onto a semi-transparent screen. While the tilt of the projector is fix, the tilt of the mirror is adjustable. By that, the redirected image can be displayed on the screen without distortion. In return, that avoids the need for additional software warping or keystone correction. All areas of the "motor compartment" that reflect light and by that, influence the image on the semi-transparent screen by reflection, are covered with black velvet. The projection screen is a semi-transparent screen (AV Vision White/Black Sunlight Screen[8]).

The design process of the dashboards was driven by two aspects: first, it should offer a large or maximized display space. Second, it should be comfortably reachable for interaction (e.g. gestures or touch). Considering the layout of today's cars, the final dashboard has an L-shaped size. It covers the area from the left of the steering wheel to the right side of the vertical center stack. On the left side, it extends to the bottom of the vertical center stack. The projection area has a maximum width of 76.5 cm and a maximum height of 68 cm. The height of the long side of the L is 28.5 cm and the width of the short side of the L is 18.5 cm. By that, the dashboard offers an overall interaction space of 2837.75 cm^2 or 0.28 m^2.

If the video signal has the correct format (side-by-side, top-bottom, or frame-alternating), the projector can be configured to convert the video signal into 3D images by processing and alternatingly presenting them. Shutter glasses that are in sync with the projector can then make sure that the correct image is only seen by the

---

[8] https://www.avdata.de/, AVData, 2020-09-23

respective eye. Synchronization is realized via an IR-emitter. In our case, we use a volfoni ActivHub RF50 emitter[9] and volfoni Edge RF active stereo glasses[10].

The projector of the buck is connected to the projector of the environment simulation to synchronize the images and prevent a possible color shift that would arise due to the shutter glasses. The projector is connected to the same workstation as the environment simulation. The buck has exhaust openings at the top and the bottom to allow air circulation and prevent overheating of the projector. All cables are wired at the bottom of the buck.

(a) Perspective view   (b) Side view

Figure 4.3 Side and perspective view of the real buck

Figure 4.3 shows the final buck in a side view with removed side panel to show the projector and mirror setup. It is only similar to the initial 3D model. There were some modifications necessary to improve design and reduce weight. First, the adjustment and mounting setup for the first-surface mirror was redesigned to have a smaller form factor. It consists only of adjusting screws that are attached to the buck and the mirror. Due to the absence of the real car seat or appropriate measures during the planning phase, the seat is mounted slightly lower than originally planned. That leads to the fact that, for smaller people (smaller than 155 cm) adjustments are necessary to allow for a comfortable driving experience. For them, pedals need to be elevated and the seat needs to be set to its highest position. In addition to that, the mount for the steering wheel has only one connection to the floor instead of two. This way, a larger area of the dashboard is visible to the driver from the default viewing position. Also, on the rear, the aluminum profiles were removed to save weight.

---

[9] http://volfoni.com/en/activhub-rf50/, volfoni ActivHub RF50, 2020-09-23

[10] http://volfoni.com/en/edge-rf/, volfoni Edge RF, 2020-09-23

To realize the off-axis projection and by that, motion parallax, an Optitrack Motive system with six Prime 13[11] cameras was installed.

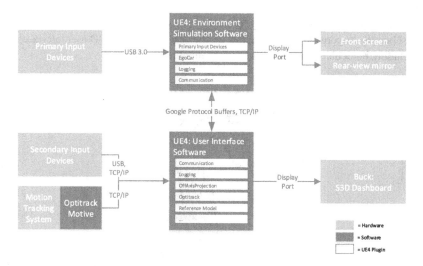

Figure 4.4 Software architecture of the prototyping setup. Faded elements are hardware components

Figure 4.4 highlights the single software components of the prototyping setup. The upper part shows the environment simulation, the lower part the user interface simulation. Optitrack Motive is a necessary software to receive and distribute the tracking data. It is used for head-tracking and by that, enables the off-axis projection but can also be used for full-body tracking. It is however, marker-based. Hence, it requires the driver (or participant) to wear items that can be tracked (e.g. glasses or gloves). We further use Optitrack Motives built-in algorithms to apply forward prediction (set to a value of 10) and for smoothing of tracking data (set to value of 20). Both values were chosen based on recommendations of the Optitrack support team.

The environment simulation software is responsible for vehicle behavior within a driving environment. Based on the review of other available software and tools at the time, we decided to rely on the built-in functionality of Unreal Engine 4. As a basis for a driving simulation and user interface software, it offers state-of-the-art

---

[11] http://www.optitrack.com, Optitrack, 2020-09-23

graphic fidelity, animated actors, a vehicle physics model, and necessary authoring functionality for experts in C++ and for novices using their visual programming approach *Blueprints*. The design was briefly presented in the previous section (Section 4.3).

The UI software is the primary tool used in this thesis to test and evaluate S3D dashboards. Its core purpose is to allow for the design of various 3D user interfaces, including binocular depth cues and various input modalities.

Figure 4.5 3D model or template used for development of the S3D dashboard

Figure 4.5 shows the 3D reference model used in UE4 to design user interface prototypes. In this figure, the black mesh-like area depicts the L-shaped dashboard and usable area. The grey cube behind it has been added as a reference to indicate depth and the upper limit of the zone of comfort. The red cube represents a tracking target that is attached to the buck. The black cubes are used to configure the off-axis projection. The green planes depict the difference between the image size of the projector and the visible area of the projection screen. We will outline these features in more details in the following paragraphs.

The user interface software has a similar architecture as the driving simulation software. It is built with Unreal Engine 4 and uses several plugins. For logging and communication, it uses instances of the same plugins as the environment simulation software. The Optitrack plugin receives data from Motive via UDP. After unpacking and deserialization, the data of all tracked objects is stored in a map. The key to the map is the identifier, the value is a data structure containing the tracking data (6 degree of freedom). The plugin offers an interface for accessing this map in UE4 and any other plugins. Access to the map is atomic. Tracking data is only updated for those objects where valid tracking data is available. Note that Motive performs interpolation and smoothing. However, it is still possible that Motive sends invalid data, for example when tracking is lost for a longer period of time due to occlusion of the tracking target. If tracking for one or more objects and hence, data is invalid, the system reuses the last valid entry. The combination of this approach — smoothing, interpolation, and reuse — ensures a rather fault-tolerant usage of tracking data.

The *OffAxisPlugin* is the primary user of the tracking data. Two main bodies are tracked for the off-axis projection: the users *head* and the *buck*. The tracking target representing the head are the 3D glasses. Participants have to wear the glasses to perceive the binocular depth cues (we apply active stereo). Attached to the glasses are tracking targets that represent the participant's head. The calculated position of this target is the geometric center of mass formed by all markers. To accurately use it as a representation of the head, a posterior offset of 12 cm is added (from the original point towards the back of the head). By that, if the participant wears glasses and rotates the head, only relatively small positional displacements of the virtual camera are visible. While a per-participant calibration procedure could remove even these movements, such a procedure would take up a significant time of the available per-participant time budget in user studies. The corrected head-position is used by UE4 and the OffAxisProjection plugin to calculate the off-axis projection matrix. To perform this calculation, the position of the second tracked entity is necessary: the buck. In detail, the off-axis projection relies on the positions of the screen's corners. In our case, where the screen corner positions are always relative to the buck — the semi-transparent screen and the projector are attached to the buck — an initial calibration procedure is sufficient to specify the screen's corners relative to a single tracking target. Hence, the buck is represented by a tracking target that is mounted on top of the hood. The red cube in Figure 4.5 illustrates the target for the buck. In reality, the buck has a physical tracking target at that very position. UE4 then takes the tracked position of the target and applies it to the dashboard. To calculate the accurate projection matrix, the corners of the projected image are necessary (c.f. black cubes in Figure 4.5). Note that the image created by the projector is slightly wider than the semi-transparent screen due to the 16:10 format. This difference is indicated by the green planes in figure Figure 4.5. Because the height of the semi-transparent screen is known and is the same as the image of the projector, calibration is a straightforward process by applying the aspect ratio of the projector's WQXGA resolution (2560×600 px, 16:10) to calculate the position of the corner points.

The positional data of the head and the screen corners are then used to calculate the new modified projection matrix: usually, the virtual camera (or the eye) is located in the center of the view frustum (c.f. Figure 4.6a). For the off-axis projection, the view frustum is asymmetric (c.f. Figure 4.6b). Hence, the first step is to calculate

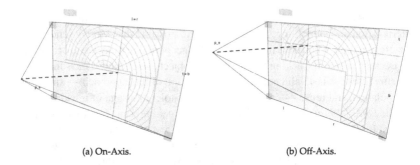

(a) On-Axis.   (b) Off-Axis.

Figure 4.6 Illustration of on-axis (Figure 4.6a) and off-axis projection (Figure 4.6b). The view frustum in the off-axis projection is asymmetric

the asymmetric view frustum based on the screens corner vectors. Next, the frustum is translated so the origin is located at the head position provided by the tracking system. For the stereo effect, a translation of half the IPD is applied along the axis formed by the two eyes. After that, the resulting matrix is normalized to fit the UE4 graphics pipeline and updated in UE4. If no off-axis projection is applied, using Optitrack is not necessary. The system then uses the on-axis projection, meaning the regular projection matrix. However, this approach introduces shear-distortion. This happens especially if the head moves.

It is important to make sure that the process of acquiring tracking data, updating the projection matrix, rendering the frame, and displaying the final image does not introduce any significant lag. Lag can lead to a unpleasant user experience [Pausch et al., 1992], performance decrease [Arthur et al., 1993; Pausch et al., 1992], and also sometimes to increased simulator sickness symptoms [Buker et al., 2012]. Lag can be described by the overall latency L of the setup. In other words, the time it takes the system to react to user input and display the according updated image. That means the data path from Optitrack's tracking cameras to the displayed image of the projector. Hence, L can be defined as follows:

$$L = L_{Motive} + L_{UE4} + L_{Projector} \qquad (4.1)$$

In our setup, $L_{Motive} = 5.18\,\text{ms}$[12]. UE4 operates at the mentioned worksta-tion with a fixed frame rate of 120 fps. That leads to a between-frame latency

---

[12] https://v21.wiki.optitrack.com/index.php?title=Latency_Measurements Client Latency, retrieved 2020-09-23

$L_{UE4} = 8.3\,\text{ms}$[13]. UE4 gathers data every frame as well as calculates and applies the modified projection matrix (the overall calculation of the projection matrix takes 0.04 ms). Finally, the projector displays images with 120 Hz. That leads to a between-image time of $L_{Projector-2.5D} = 8.3\,\text{ms}$ per image. With S3D, refresh rate per eye is half of that: $L_{Projector-S3D} = 16.6\,\text{ms}$. That leads to an overall latency estimation of $L = 5.18\,\text{ms} + 8.3\,\text{ms} + 16.6\,\text{ms} = 30.08\,\text{ms}$. Note that this calculation is only an informed estimation. It does not include transmission between devices (e.g. workstation and projector). Hence, the absolute latency L is slightly higher. For an accurate motion-to-photon latency measurement, special equipment is necessary [Choi et al., 2018]. All measurements were acquired with the driving simulation demo scene shown in Figure 4.1 running. Data has been measured for five minutes and the arithmetic average has been reported.

4.5 Discussion

The last two sections presented the prototyping setup that has been developed over the course of this research. This section will discuss its advantages as well as possible improvements. The setup consists of two general parts:

1. Environment simulation: driving simulation software as well as audio and video hardware.
2. The buck: car mock-up with S3D dashboard (rear-projection) and user interface prototyping tools.

Designing and building this environment from scratch led to a flexible, and easily reconfigurable environment. It has been used successfully for several user studies in the area of S3D dashboards — some of them are part of this thesis. It has also been used for user studies that were of more general nature and where S3D did not play a vital role. For example, in Hoesch et al. [2016], we compared the impact of various types of environment simulations (large projection screen in 2.5D, large projection screen in 3D, and VR-HMD) on simulator sickness, bio-physiological measures, and driving performance. In another example, Hoesch et al. [2018] investigated the relationship between visual attention and simulator sickness using the setup.

Initially, the environment simulation did not have a rear-view. while this was appropriate for studies without any cars (e.g. the Lane Change Task [Mattes, 2003] used in Hoesch et al. [2016]), a rear-view was necessary in later studies. Here,

---

[13] Triple Buffering could have increased this value but was disabled.

selecting UE4 because of its easy extensibility and flexible nature proofed to be beneficial. Relying on the multiplayer mechanism of UE4 allowed us to build a client-server application that could then be used as a rear view. To do this, two applications were started — one on the workstation driving the simulation and one at a dedicated workstation that was connected to a second projector displaying the rear-view behind the buck. The same approach — having a client application and a separate display — can be used for side mirrors in the future. As an alternative and if the rear projection provides a large enough view, real mirrors can be attached to the buck and provide a rear view.

4.5.1 Simulator Fidelity

The integrated projectors are able to display images using a high refresh rate of 120 Hz and a WQXGA resolution. In combination with the connected workstation, the system is able to provide a rather low system latency. However, the system has some limitations that need to be considered when interpreting results but also when designing user studies. The buck currently lacks accurate side-view mirrors (a rear-view mirror was applied for a more natural driving environment). Adding real side-view mirrors would have further enhanced the environment but was not possible due to laboratory restrictions on available hardware. Also, there was no seat-belt. Naturally, there was no risk of a crash in our system but movement patterns of participants in cars are usually restricted by the seat belt. Hence, our setup allowed more free movement which need to be considered for some studies (c.f. Chapter 9). Also, the buck can be extended to form the whole front of the car. The system was designed in such a way that a mirrored version of the buck can be attached to the right side (without primary input devices). The aluminum frame can be replaced, and the projector can be located in the center of the combined buck. By that, the projector is able to augment two L-shaped displays (or one large T-shaped display) and provide an even larger S3D dashboard, even for a passenger. It is important to note that the absence of a motion platform limits our results. While the design of the buck including the rear-projection system would have allowed mounting the buck on a motion platform, this was not feasible within this research. Further, for our prototyping setup, using a real car or part of a real car as used in other simulators (c.f. Gabbard et al. [2019]) was not suitable because of the S3D dashboard. Integrating it into an existing car layout would have required extensive work that would have been inappropriate, given the experimental nature of the research. Hence, we decided to build a mock-up that is similar to a real car in dimension and allows for rear-projection.

At the time, VR-HMDs with sufficient resolution and refresh rate were not available so a non-HMD based solution was designed. With VR-HMDs, the human body is often not visible (or not accurately represented) which can lead to a reduced sense of embodiment [Kilteni et al., 2012; Perez-Marcos et al., 2012], and might hamper interaction with input devices. However, both, the driving simulation and the user interface can be used with state-of-the-art VR-HMDs due to Unreal Engines support (c.f. Weidner et al. (2017)). If this is the case, the buck can used for passive haptic feedback to increase presence by providing haptic cues to the driver (e.g. for the steering wheel, the dashboard, or the sifter).

Overall, the setup provides a medium-fidelity driving simulator, focusing on high-resolution and high refresh rate, while conveying the layout of an in-car environment.

4.5.2   Graphics Fidelity

The spatial AR arrangement creating our binocular depth cues on the dashboard provides a good graphics fidelity and integrates well with the first-surface mirror and the semi-transparent screen. Other (and future) versions of such prototypes should especially pay attention to an even and well distributed gain curve of the semi-transparent screen. In our case, brightness of the screen at higher viewing angles is visibly lower than with viewing angles where the viewing direction is orthogonal to the screen. Hence, a semi-transparent screen with a close-to-linear gain curve should be used. However, for our setup — while it required additional considerations when designing UIs, e.g. making items located in outer areas brighter— the gain curve was sufficient. A further inherent limitation of this type of AR is that the glasses do not cover the whole field of view. Parts at the outer edges of the field of view are either obstructed by the frame of the 3D glasses or not covered by them (the 3D effect is only visible when looking through the glasses, otherwise double images are visible). However, when looking through the glasses, the 3D effect promised to be better than with autostereoscopic displays that were available at the time. Novel displays promise to provide an equally good experience without the need for glasses and could be viable candidates for future research (e.g. lightfield displays[14] or high-resolution autostereoscopic displays[15]).

---

[14] https://www.leiainc.com/, Leia Inc., 2020-09-23

[15] https://lookingglassfactory.com/product/8k, Looking Glass Factory — 8k immersive display, 2020-09-23

Choosing Unreal Engine 4 as basis for the driving simulation has allowed for a streamlined development of the driving simulation. It provides full source code access, has a good documentation, periodic release cycles and a huge community. In addition to a good level editor, it provides sophisticated rendering algorithms. Both speed up the development process significantly. However, the periodic releases also introduce additional work to keep software up-to-date. While providing a good visualization, there are some drawbacks using the Unreal Engine 4 for the driving simulation: first, the provided vehicle physics model is far from realistic, necessitating intensive tuning of parameters. It also changed several times during release cycles with made it necessary to readjust parameters. Second, there is currently no automated system for moving or simulating of traffic actors. That makes it hard to create sophisticated urban scenarios with many traffic participants without modeling movement and behavior of individual actors manually. In our driving simulation scenarios, we had to do exactly that in a rather inefficient process. However, the experiments did not necessitate many other actors or complex maneuvers. The integration of a system like SUMO[16], a traffic simulator, could solve this problem. The UE4 artificial intelligence system[17] could be another viable candidate for future improvements regarding traffic simulation.

4.5.3 Requirements

After having outlined the advantages but also flaws of our system, the next paragraphs are going to discuss our solution with respect to the requirements formulated in Section 4.1.

Large S3D Dashboard. The setup allows us to test and develop novel UI prototypes for the interaction with stereoscopic 3D dashboards. Because the semi-transparent screen covers a large area of the dashboard on the driver's side, the design space is significantly larger than with conventional dashboards and also larger than in previous research that applied binocular depth cues on the dashboard.

Graphic Fidelity. The projection hardware of the simulation environment provides a sufficient level of detail with a pixel size of 1.4 mm × 1.4 mm with a refresh rate of 120 Hz. The center of the buck's seat is located 3 m in front of the screen which

---

[16] http://sumo.dlr.de/, SUMO — Simulation of Urban Mobility, 2020-09-23

[17] https://docs.unrealengine.com/en-US/Engine/ArtificialIntelligence/, UE4 Artificial Intelligence System, 2020-09-23

makes differences between pixels not noticeable. The average pixel size on the S3D dashboard is 0.4 mm × 0.4 mm. Using UE4 further provides the setup for with the state-of-the-art rendering algorithms that can be used for the driving simulation as well as the user interface. However the gain-curve of the semi-transparent screen influenced results.

Interactivity. The primary interaction devices of the buck are able to provide a plausible feeling during driving. The force-feedback of the steering wheel adds to the simulator fidelity. However, the gears' teeth (overall 900) of the internal step motor are noticeable and sometimes prevent precise steering. In UE4, force feed-back can be controlled and adjusted to fit the car, the environment, and events. For secondary input devices, UE4 provides a convenient plugin system that can be used to add devices (e.g. LeapMotion[18] or Microsoft Kinect[19]) Further, interaction via motion tracking systems like Optitrack is possible. Integrating these devices to cre-ate functional prototypes is possible by two ways: for experts and for high level of customization, C++ can be used to add new functionality. To use existing function-ality, and by that, by novices, UE4's visual programming approach *Blueprints* can be used. The latter has been shown to be easier than regular text-based coding (like, e.g. in Unity) for novices in game design [Dickson et al., 2017].

Reconfiguration. The buck is designed to be rather modular. Besides the dynamic placement and arrangement of the primary interaction devices and the possibility to add secondary input hardware, its design based on aluminum profiles allows for modifications. The plugin-based approach for the driving simulation and the user interface further support re-configurations and varying setups. This and the network based approach further enables the usage of the driving simulation with other user interface applications and vice versa (c.f. Weidner et al. [2019]).

Rapid Prototyping The hardware and software has been designed to enable adding or removing other devices or parts by means of a standardized systems and accessible aluminum profiles. More important, the software allows for easy and fast develop-ment cycles with 3D models and interaction artifacts that have increasingly better fidelity. The plugin-based approach as well as the separation of driving simulation and user interface software further allow for reuse but also concurrent development in teams. For low-fidelity prototyping, the semi-transparent screen can easily be exchanged with a L-shaped chalk-board. This allows for the rapid design of user

---

[18] http://www.ultraleap.com, Leap Motion, 2020-09-23

[19] https://developer.microsoft.com/de-de/windows/kinect, Kinect for Windows, 2020-09-23

interface, e.g. in a participatory design process. Here, the depth of elements can be indicated by magnetic building blocks for negative disparity and by drawings for positive disparity (c.f [Haupt et al., 2018]).

Connectivity Connectivity has been established by several mechanisms. First, we divided the user interface software and the driving simulation software. They exchange data via local area network and are otherwise decoupled. Also, the driving simulation software can be used in multiplayer mode for a distributed driving simulation but also for additional views (e.g. the rear-view mirror). The environment further allows for the integration of other tools and services that are based on communication. For example, the integration of a Wizard-Of-Oz tools that control the user interface and realize an interactive speech-based interaction is easily possible (c.f. Weidner et al. [2019]).

In summary, the developed driving simulation environment fulfills the initially posed requirements. It offers a flexible, extensible, and easy-to-use environment for the development of prototypes for the exploration of S3D dashboards in vehicles.

Design Space for S3D Dashboards

5

> Parts of this section are based on Weidner and Broll [2017b].

5.1   Problem Statement

When designing and evaluating S3D UIs, it is necessary to ensure that all participants can fuse the displayed stereoscopic image pairs. By that, they can perceive the full UI and the depth cues without loss of fusion or binocular rivalry. However, as mentioned in Section 2.1 and Section 2.2, the overall ability to fuse S3D images varies between humans. Every user has a personal zone of comfort, meaning that every person has a maximum and minimum image disparity that can be comfortably fused. That means that — while display systems could produce images with very large disparities and by that, convey very large distances — the human visual system can only process them within a certain boundary. Displaying content outside the zone of comfort can lead to double vision and/or visual fatigue. That, in return, can impact user experience and safety.

Considering the fact that the display in the buck attracts glances with non-orthogonal and large viewing angles and that it, due to using a projector, leads to slightly irregular light distribution, we evaluate the zone of comfort for the previously mentioned design. By that, we define the effective volume for S3D elements and S3D UIs which we will use in future studies.

Electronic supplementary material The online version of this chapter (https://doi.org/10.1007/978-3-658-35147-2_5) contains supplementary material, which is available to authorized users.

F. Weidner, *S3D Dashboard*,
https://doi.org/10.1007/978-3-658-35147-2_5

5.2   On the Design of S3D UIs Inside and Outside the cars

Earlier research suggests that disparity should not exceed values between -25% and +60% of the viewing distance [Williams and Parrish, 1990]. For our simulator and an approximate distance between the drivers head and the screen of 80 cm, that would result in a depth budget along the line-of-sight of -20 cm (towards the user) to 48 cm (away from the user). Another rule of thumb states that the minimum and maximum disparity should be within $D = \pm 1°$ [Kooi et al., 2010]. Research of Brov et al. [2013] investigated the zone of comfort for a vehicle instrument cluster. They recommend using disparity values not larger (or lower) than $D = \pm 35.9$ arc-min which corresponds to $D = \pm 0.6°$. An overview of other (but similar) volume definitions has been compiled by McIntire et al. [2015]. It is noticeable that the recommendations vary. They have also been derived for orthogonal viewing angles. Also, they usually neglect display- and environment-specific aspects like display technology which influence binocular fusion. Hence, to base our results on a solid foundation and also inform future designs with large S3D dashboards, a dedicated zone of comfort is derived.

5.3   Study Design

5.3.1   Apparatus

Figure 5.1 illustrates the hard- and software setup of the system used for establishing a zone of comfort. Contrary to the description of the car mock-up in Chapter 4, head-tracking was not yet applied in this experiment. Hence, head movements have been ignored by the system. The location of the virtual camera was fixed and located 80

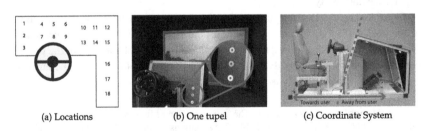

(a) Locations   (b) One tupel   (c) Coordinate System

Figure 5.1 Setup of the experiment determining the zone of comfort. Figure 5.1a shows the locations where the visual stimulus, shown in Figure 5.1b, appears. Figure 5.1c illustrates the directions towards and away from the user

cm in front of the dashboard along an orthogonal axis going through the middle of the dashboard. Also, the borders of the aluminum profiles facing the user were not coated in black velvet yet. Hence, the aluminum profiles are noticeable. The user interface software has been implemented using Unreal Engine 4.16. With this setup, the zone of comfort has been derived.

5.3.2   Procedure

After participants arrived at the laboratory, they were asked to fill out a questionnaire asking for demographics (c.f. electronic supplementary material B4). Next, they had to pass a Random Dot Stereogram [Schwartz and Krantz, 2018] to confirm their general capability of binocular fusion. Then they took a seat in the buck, adjusted the seat, and put on glasses. The task of participants was to adjust the disparity D of one of three tori. An exemplary tuple is shown in Figure 5.1b. One torus had the size of 4 cm. Space between two tori was 5.5 cm. The three tori appeared on several locations on the dashboard as indicated by Figure 5.1a, starting with locations 1, 2, and 3. Participants then used buttons on the steering wheel to change disparity. The buttons moved the active torus 1 mm towards and away from the user and by that, changed disparity and perceived depth. Note that the movement was towards the position of the participant's head. As mentioned before, the head position was fixed at the position of the virtual human's head illustrated in the CAD-model shown in Figure 4.6a. Participants adjusted the disparity for one torus until they either lost fusion or perceived it as uncomfortable. They then confirmed this position and selected the next torus of the triple with buttons on the steering wheel. This also reset the disparity of the adjusted torus. This way, there were always two tori acting as a reference with $D = 0°$ (as suggested by Broy et al. [2013]) whereas one was adjusted. After having adjusted all tori of the triple, the application went to the next triple. Finally, after having adjusted all tori of all triples towards one direction (either towards or away as indicated by Figure 5.1c), they were asked to adjust the triples in the opposite direction. The order of direction was counterbalanced.

Overall, There were 18 tori (6 triple á 3 tori). The locations were not uniformly distributed but spaced in such a way that the locations cover the relevant dashboard areas mentioned by Kern and Schmidt [2009] but also so that the experiment duration is not unnecessarily prolonged. Shape and color of the objects was inspired by Pitts et al. [2015] who performed similar perception experiments with in-car S3D.

5.3.3 Sample

25 participants were invited to participate in the study (male = 20, female = 5, mean age = 25.8 ± 4.3 years, min = 18 years, max = 36 years). All had normal or corrected to normal vision and passed a stereo test [Schwartz and Krantz, Schwartz and Krantz 2018]. The invitation was distributed via university mailing list and university Facebook groups. Hence, participants were mostly students and university staff. Note, this study was performed in conjunction with the study investigating driver distraction (c.f. Chapter 6).

5.4   Results

Figure 5.2 illustrates the zone of comfort. Figure 5.2a shows the closest locations in cm where virtual content should be located. Similarly, Figure 5.2b shows the farthest location where virtual content could be located. For example, 36 in Figure 5.2b means that virtual elements should not be more than 36 cm away from the projection screen. If content is located outside these boundaries, it is likely that users will perceive double images or visual fatigue. Numbers written in bold represent points on the dashboard where measurements were taken. Numbers not written in bold have been calculated using a 3×3 box filter. The detailed measurements including disparity values for a viewing distance of 80 cm are listed in the electronic supplementary material A.

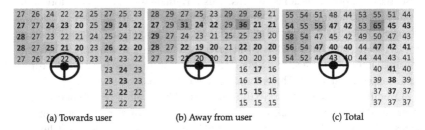

(a) Towards user   (b) Away from user   (c) Total

Figure 5.2 Zone of comfort in cm. Figure 5.2a and Figure 5.2b indicate the extends starting from the semi-transparent screen towards and away from the user in cm. Figure 5.2c shows the combined zone of comfort in cm. The more saturated the green, the larger the zone of comfort. Bold numbers represent measurement points whereas non-bold numbers have been interpolated using a 3 × 3 box filter

5.5   Discussion

The measured zone of comfort offers a rather large area where S3D content can be displayed. However, there are several aspects of the measured zone of comfort worth discussing. Prior work suggests that the zone of comfort is ideal when looking orthogonal at an image pair that is located on a vertical and horizontal horopter. However, this ignores factors introduced by display technologies and simulator setup. Hence, our measured zone of comfort deviates from this perfect scenario and participants reported slightly different values. In the following, the outline of the zone of comfort is discussed as well as some salient areas.

For the upper part of the dashboard, the zone of comfort shows similar characteristics for both directions, towards and away from the user. Slightly higher on the left side, with degradations towards the right side. Both sides show a slight drop above and to the right of the steering wheel. This symmetry was expected and confirms with previous results [Howard and Rogers, 2008].

There is a slight decline of the effective range above the steering wheel. This is most likely due to the projector and its arrangement. The projector is located in the fake motor compartment and the image is redirected by a mirror. Because the gain curve of the semi-transparent screen is not perfect, a hot spot is visible at this location. Looking at this hot spot, a brighter image is visible. While this hot spot could be eliminated by reducing the overall brightness, this would have degraded the brightness in the outer areas of the dashboard. Hence, when using such a setup with projector and semi-transparent screen, careful adjustment of the brightness is necessary.

Finally, the zone of comfort decreases towards the right side of the dashboard. This is most likely a result of the high viewing angle with respect to the projection plane. Similarly, the lower vertical center console shows a smaller zone of comfort. Especially the away-condition shows a smaller effective range. One possible explanation for this could be that participants' eyes accommodated at the screen but vergence was adjusted to look past the semi-transparent screen and were disturbed by the aluminum profiles. While the interior of the motor compartment was dark, the uncovered aluminum parts were slightly visible. Hence, they could have impaired focus and binocular fusion.

These results act as a guidance for S3D dashboard designs and offer boundaries on where to locate content. Comparing our results with other rules of thumb (e.g. 1°-degree-rule, [Lambooij et al., Lambooij et al. 2009], our results often indicate a depth budget higher than 1° in both directions, sometimes reaching values as high as 2.04° . For practical purposes, that would not pose a problem because the 1° rule would still ensure that users are able to fuse the double images — however it would restrict the available design space. More important, results also indicate locations where the available depth budget is smaller than $1°$. Relying on the rule-of-thumb from the literature would mean that it is likely that participants are not being able to fuse the stereoscopic image pairs and by that experience double vision and visual discomfort. Also, in the center at the instrument cluster (locations 4–6), Broy et al. [2013] have derived a zone of comfort which is with $\pm 0.6°$ slightly more conservative than our zone of comfort at this location (appr. $\pm 1.06°$).

The various recommendations make it necessary to consider that the derived zone of comfort is valid for our simulator and can only act as an orientation for other setups — it does not make pretests unnecessary. For example, the objects to be manipulated by participants were without color and had the same contrast levels. However, as mentioned in Section 2.1, color but also other factors influence binocular fusion. Also, the sample was rather small and limited to a narrow age range. However, given the comparison with other recommendations, it seems to be important to measure the zone of comfort for specific applications to find weak spots in the setup, that arise from the displays or viewing angle.

With the volume described by the measurement points, a defined volume for S3D content is available. This region can now be used to design and develop S3D UIs. All S3D content should not use disparity values outside this region.

In conclusion, the study offers a defined volume where virtual objects can be located. The study was necessary to detect and deal with the imperfections of the display setup. They manifest themselves in two specific properties of the zone of comfort: the narrower zone of comfort at the instrument cluster where the brightness hot spot is located and the smaller volume at the lower right center console where the viewing angle on the semi-transparent screen is far from being orthogonal. Using this derived zone of comfort, designers and developers are now able to build dashboard prototypes that are likely to contain no images that cannot be fused by the human visual system and that allows them to respect the specific properties of the setup.

Part III
Case Studies & Applications

This part portrays exemplary case studies and applications of S3D dashboards. Based on the analysis in Chapter 2 and Chapter 3, we present user studies exploring S3D dashboards in the area of driver distraction, trust, take-over maneuvers, and derive interaction techniques for the interaction with S3D dashboards.

S3D for Change Detection and System Control

6

Parts of this section are based on Weidner and Broll [2019a].

6.1 Problem Statement

When driving manually, people often have to react to stimuli on the dashboard, for example flashing warning signs on the instrument cluster. Also, many applications in the vehicle are structured as text-based, hierarchical menus. For example, the entertainment or the navigation system. When executing such tasks, the user is distracted and cannot fully focus on the road and traffic. With distracted driving being one of the main causes for accidents, these operations pose a safety risk. User interfaces should support the driver during these maneuvers by prominently highlighting important elements (e.g. notifications or a selected item). Thus, we use these tasks—detecting changes and navigating list-based menus—to get insights on the performance of S3D dashboards.

Chapter 3 described several studies on how S3D improved menu navigation [Broy et al., 2012] and detecting changes on the user interface [Szczerba and Hersberger, 2014]. However, these studies have been performed without a primary driving task. By that, participants could fully focus on the task that involved S3D. However, in a vehicle, driving is often the primary task whereas the manipulation of content is the secondary task. This requires sharing mental resources and might impact per-

---

Electronic supplementary material The online version of this chapter (https://doi.org/10.1007/978-3-658-35147-2_6) contains supplementary material, which is available to authorized users.

formance [Wickens, 2002]. Also, all experiments were performed in a setup where the S3D display was located behind the steering wheel, substituting the instrument cluster. Other possible display locations where content is displayed in traditional cars and where it might be displayed in the future, have not been investigated yet.

To evaluate whether an S3D dashboard has the potential to increase these two aspects of driving (reacting to stimuli and navigating list-based menus), a user study was executed that compared 2.5D and 3D visualizations. Considering the related work of S3D (positive results in task performance but also increased simulator sickness), we anticipate two possible outcomes of this experiment: On the one hand, it could be that S3D enhances performance due to binocular disparity and the pop-out effect. On the other hand, it could be that S3D is perceived as a distraction and impairs task performance. Hence, the overall research questions are as follows:

RQ1:   Can S3D support these two types of tasks while driving and by that, improve primary and secondary task performance?

RQ2:   Does the location of the S3D content influence task performance?

Primary task is the driving task, whereas the secondary tasks in this experiment are change detection and list selection. Based on this, the following research hypothesis were formulated:

H1:   S3D influences reaction time during change detection.
H2:   S3D influences the number of changes detected correctly.
H3:   S3D influences task completion time during list selection.
H4:   S3D influences number of lists completed correctly.
H5:   S3D influences performance in the car following task.
H6:   S3D increases simulator sickness.

6.2   Study Design

We used a between-subject design for this user study. Between-subject factor was the display modality (2.5D vs S3D). Each participant had to perform two secondary tasks (change detection and list selection) while performing the primary task (following a car while driving manually). Participants started with the change detection task and then performed the list selection task. Note, this study was performed in conjunction with the study determining the zone of comfort (c.f. Chapter 5).

6.2.1 Apparatus

The simulator used for this study was similar to the one described in Chapter 5 but the borders of the frame were not coated yet. Also, neither head-tracking, nor a rear-view mirror was applied.

6.2.2 Procedure

Participants had to complete two approximately 12 minute long drives while doing the primary task. The primary task was following a car and keeping the distance to it within a certain range (car following task). There were no lane markings and participants were encouraged to follow general traffic laws. In addition to the primary task, participants had to perform one secondary task per drive. First a change detection task, then a list selection task. Because we do not compare the results of the primary tasks with each other, order effects can be neglected [Elmes et al., 2012]. The following sections are going to explain the tasks in detail.

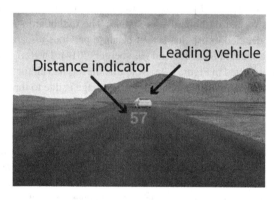

Figure 6.1 Screenshot of the driving environment showing the leading car, and the green distance indicator. The indicator turns red if the distance is above 120 m or below 40 m

Primary Task: Car Following Task
Participants had to drive along a simple rectangular course (one lane, no contra-flow) and follow a leading car (c.f. electronic supplementary material B1). The ego-car was set up to automatically change gears. Participants were told to keep the distance to the leading car between 40 m and 120 m. Figure 6.1 shows a screenshot of the

driving environment. Because we wanted participants to look at the road—to prevent them focusing on the dashboard and the displayed content—a distance indicator was placed on the front screen, simulating a simple HUD display. The indicator showed the distance in meters and changed color according to the distance. It was green when the distance was between 40 m and 120 m and turned red when the distance was greater than 120 m or smaller than 40 m. The upper interval of 120 m was added to prevent participants from simply letting the leading vehicle drive away and by that, avoiding the primary task.

For the two secondary tasks, the mean speed of the leading car was different. The leading car had an average speed of 90 km/h during the change detection task. During list selection, the car drove 70 km/h. In both scenarios, it decelerated slightly when taking curves. This is because of various reasons. First, we wanted to allude a training effect in the driving task, hence a different speed was chosen. Second, adjusting the speed allowed us to re-use the track and place an appropriate number of events (changes to detect and lists to select) in it without any major redesigns of the track.

Secondary Task 1: Change Detection
This secondary task required participants to press a button on the steering wheel when a stimulus on the dashboard was visible. The dashboard showed a matrix of grey tori. We decided to use this layout because previous research used similar designs [Pitts et al., 2015; Szczerba and Hersberger, 2014]. The tori were displayed at the locations indicated in Figure 6.2a. Note that the tori had all the same color and contrast for the user. The uneven contrast in the image is due to the viewing angle and the camera equipment.

Every 10–20 s, a change happened. That means that, depending on the condition, one torus either increased size (2.5D) or moved towards the participant (S3D). After two seconds, the torus returned to its original size or location. Participants were required to press a button during these 2 s. Note that the only difference between the two visualizations are the binocular depth cues. For the S3D condition, initial disparity of each torus was $D = 0°$. During the change, disparity was changed to $D = -0.95°$ (13.6 cm towards the user). The torus' position was changed along the axis between its initial position and the virtual camera. Note that this study was performed in parallel with and before the one that determined the comfort zone. Hence, the design conservatively followed the 1°-rule. However, this also means that the change was sometimes larger than the later measured zone of comfort which needs to be considered when discussing the results.

(a) No change.   (b) Change present.

Figure 6.2 Illustration of the dashboard layout used for the change detection task. Every 10–20 s a change happened. That means, one of the tori increased size or moved towards the user for 2 s before it returned to its original state

There were 58 changes in total. They happened in randomized order for each participant, but it was ensured that each torus at each location changed twice. The first four changes were considered training and the same for all participants. They happened on each dashboard area (c.f. Figure 6.3b, 1 – left of the steering wheel, Left-Wheel; 2 – instrument cluster, IC; 3 – upper vertical center stack, CS-high; 4 – lower vertical center stack, CS-low).

Secondary Task 2: List Selection

The second secondary task was list selection. An exemplary menu is shown in Figure 6.3b. Every 20 s, a target string appeared on the front screen. Participants were then required to select the target string in a menu that had appeared at the same time on the dashboard. It used the same color scheme as the change detection task. Each target string consisted of three words. A menu had three hierarchies and each hierarchy had five entries. The first word of the target string was in the first hierarchy of the menu. The second word in the second hierarchy and the third word in the last hierarchy. Only three menu items were visible at a time, the other two were reachable via scrolling. Scrolling as well as navigating the hierarchies (select and return) were done by dedicated buttons on the steering wheel (one button for going up, another button for going down). This task uses the same locations as the change detection task (c.f. Figure 6.3b)

| (a) Target string. | (b) Menu on instrument. |

Figure 6.3 List selection task. Figure 6.3a shows and exemplary target string that appeared on the front screen. Figure 6.3b shows a scrollable menu on the instrument cluster with 5 items (three visible at a time). The other menu locations are indicated by the dashed lines: 1 = Left of the steering wheel (Left-Wheel), 2 = instrument cluster (IC), 3 = upper vertical center stack (CS-high), 4 = lower vertical center stack (CS-low)

Overall, each participant had to complete 24 menus. Similar to the change detection task, the first four menus were considered training (one for each dashboard area). The order of the training menus was the same for all participants. The other 20 menus appeared at pseudo-random locations (each location 5 times; out of order). The target strings were the same for all participants.

The currently selected menu item (initially the first menu item) had a disparity $D = -0.95°$, meaning that it appeared closer to the user. The other menu items had a disparity $D = 0°$. For the 2.5D condition, we again changed the relative size (13.6 cm towards the user). By that, the only difference between the conditions was the binocular depth cue.

By using abstract list items and target strings, we wanted to narrow down the factors that influence results. This way, no participant was able to use prior knowledge of in-vehicle menus, music titles/artists, or addresses. By that, we hoped that we can extract more fundamental information from results on the effect of S3D on the basic workflow of selecting items in a list-based menu while driving.

Showing the target string on the front screen rather than telling participants to just hold a safe distance was done so that they had to switch between the dashboard and the front screen while selecting the menu entries. By that, we intended to trigger many switches between the two displays and by that, invoke more than one binocular fusion process.

After each drive, participants filled out the simulator sickness questionnaire.

6.2.3   Sample

61 people participated in the study (female: 29.5%, male: 68.9%, other: 1.6%; age: 24.7 ± 3.9 years; min = 18 years; max = 36 years). Participants were recruited using convenience sampling via university mailing list, Facebook groups, and direct contact. All participants had a valid driver license from their country. All had normal or corrected to normal vision and passed the stereopsis test [Schwartz and Krantz, 2018]. 80.3% have previously experienced 3D displays. 19.7% have no experience with 3D displays. 47.5% have experience with driving simulators (41.4% once, 34.5% more than once and less than five times, 24.1% more than five times). There were no dropouts. Participants had the chance to win 50 Euro. Average experiment duration was 35 minutes.

6.2.4   Measures

For the primary task, we measured how well participants kept the distance to the leading car. In detail, we measured how long participants kept the distance within the target interval and vice versa. For the change detection task, we measured reaction time (the time it took a participant to detect a change and press the button on the steering wheel) and the number of correct responses. A response was correct if the participant pushed the button on the steering wheel within 2 s after the change happened. For the list selection task, we measured task completion time (how long it took participants to select the three menu items) and how many of the 20 lists they completed correctly. We further measured simulator sickness after each of the two drives using the simulator sickness questionnaire (SSQ, [Kennedy et al., 1993]) as a control variable.

6.3   Results

Data was analyzed using Python 3.73–64 bit, statsmodels-0.11.0, and scipy-1.4.1. An α-value of 0.05 was used as significance criterion when necessary. We checked for normality of the residuals using Shapiro-Wilk tests and inspected QQ-Plots.

Homoscedasticity was tested using Levene's test [Brown and Forsythe, 1974]. Data was analyzed using Mann-Whitney-U tests and, if normality assumption was not violated, with Welch-tests.

6.3.1   Experiment 1—Change Detection

Across all participants, there were $(6 + 6 + 9 + 6) \times 2 \times 2 \times 61 = 6588$ changes (6 or 9 per area, every torus twice, 2 conditions, 61 participants).

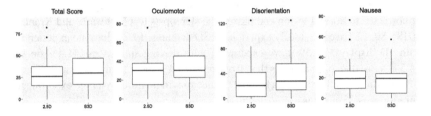

Figure 6.4 Results of SSQ after change detection task (lower is better; N [0;295.74], O [0;159.18], D [0;292.32], TS [0;235.62])

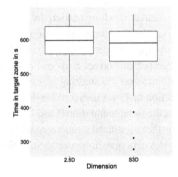

Figure 6.5 Results of the car following task during change detection (min = 0, max = 720, higher is better)

Simulator Sickness

Figure 6.4 shows the results of the SSQ after performing the change detection task. Data did not follow normal distribution. Mann-Whitney-U tests ($N1 = 30$, $N2 = 31$, one-tailed) did not indicate significant differences between the two distributions (sub-scale or total score, p > 0.05, c.f. electronic supplementary material B2). Note that we hypothesized that the S3D condition leads to more simulator sickness and hence used a one-tailed test. That tells us that it is likely that the groups did not experience different amounts of simulator sickness.

Primary Task Performance

Figure 6.5 shows information about driving performance during the change detection task. Data for was not approximately normal distributed (Shapiro-Wilk test). A Mann-Whitney-U test ($N1 = 30$, $N2 = 31$, two-tailed) showed no significant differences between distributions, suggesting that participants of the groups drove equally good or bad with respect to the car following task ($U = 530, Z = = -1.307$, $p = 0.096, r = -0.167$).

Secondary Task Performance

Task performance in the change detection task can be divided into two measures: first, reaction time to confirm the change and second, the number of changes detected correctly.

Figure 6.6 shows the results of the reaction time analysis. Normality tests indicated that data did not follow a normal distribution. Mann-Whitney-U tests ($N1 = 30$, $N2 = 31$, two-tailed) revealed no significant differences between the distributions, neither for a sub-area, nor for the total dashboard ($p > 0.05$ for all comparisons). That tells us that location or type of visualization did not impair or improved the time it took participants to push the button on the steering wheel when they detected a change on the dashboard. Complete statistical results can be found in the electronic supplementary material B2.

In Figure 6.7, results of the task performance analysis' second part are depicted: the number of changes that have been detected within 2 s. Analysis was done for each area and for the overall dashboard. After checking for normality with a Shapiro-Wilk test, which stated non-normality, Mann-Whitney-U tests ($N1 = 30$, $N2 = 31$, two-tailed) were performed. No significant differences in the distributions of the two groups were detected ($p > 0.05$ for all tests). In other words, the location or the type of visualization did not influence the likelihood of participants detecting a change on the dashboard. Complete statistical analysis can be found in the electronic supplementary material B2.

6.3.2 Experiment 2—List Selection

In total, there were $(5 + 5 + 5 + 5) \times 2 \times 61 = 2440$ lists (5 per area, 2 per condition, 61 participants).

Simulator Sickness

Data from the SSQ was not approximately normal distributed (Shapiro-Wilk test). A Mann-Whitney-U test ($N1 = 30$, $N2 = 31$, one-tailed) for two independent samples did not indicate significant differences ($p > 0.05$) between the distributions of the two groups (c.f. electronic supplementary material B3). That indicates that type of visualization and location did not impact the severity of simulator sickness symptoms (Figure 6.8).

Primary Task Performance

Figure 6.9 shows the aggregated data about driving performance. Data was not approximately normal distributed (Shapiro-Wilk test). Mann-Whitney-U tests ($N1 = 30$, $N2 = 31$, two-tailed) did not detect significant differences between the distributions ($U = 427$, $Z = 0.224$, $p = 0.589$, $r = 0.029$). That tells us that participants of each group showed a similar performance in keeping the distance to the leading vehicle within a certain range. Note that, again, we hypothesized that the S3D condition leads to more simulator sickness and hence used a one-tailed test.

Figure 6.6 Results of the reaction times during change detection task by area in ms (min = 0 ms, max = 2000 ms, lower is better)

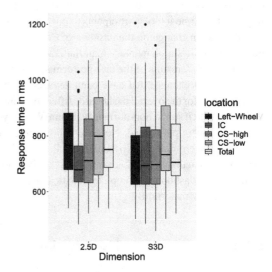

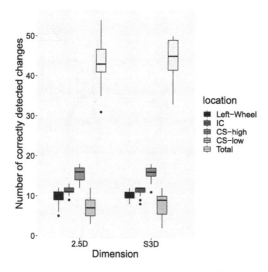

Figure 6.7 Results of changes detected correctly by area (Left-Wheel, IC, CS-High: min = 0, max = 12, CS-low: min = 0, max = 18, Total: min = 0, max = 54, higher is better)

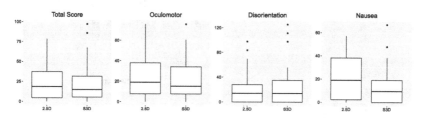

Figure 6.8 Results of the SSQ after the list selection task (N [0;295.74], O [0;159.18], D [0;292.32], TS [0;235.62], lower is better)

Secondary Task Performance

Task performance during list selection can be divided into two measures: first, task completion time to correctly select the presented word combination. Second, the number of correctly selected combinations.

Figure 6.10 shows the results of the task completion time. Normality tests did not confirm that data follows an approximation of the normal distribution. Mann-Whitney-U tests for two independent samples showed no significant differences between groups ($N1 = 30$, $N2 = 31$, two-tailed, p > 0.05). Neither for specific areas, nor for the overall dashboard. That tells us that the S3D effect and the type

Figure 6.9 Results of the
car following task during
list selection (d = median
distance [m]; in = inside
target distance interval [s];
$in_{min} = 0$, $in_{max} = 720$,
higher is better)

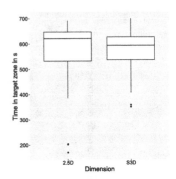

Figure 6.10 Results of task
completion times during list
selection for correct lists by
area (lower is better)

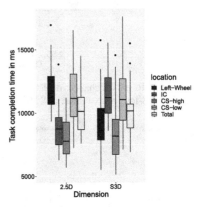

Figure 6.11 Results of the
total number of lists
selected correctly by area
(area: min = 0, max = 5;
total: min = 0 , max = 20,
higher is better)

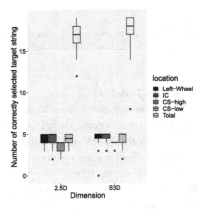
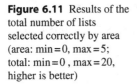

of visualization did not impact the time it took participants to successfully select the target strings in the hierarchical list. Results can be found in the electronic supplementary material B3.

Data representing the number of correct lists was also not approximately normal distributed. Mann-Whitney-U tests ($N1 = 30$, $N2 = 31$ two-tailed) for two independent samples did not indicate significant differences between the two distributions ($p > 0.5$; c.f. Figure 6.11 and the electronic supplementary material B.3). In other words, neither type of visualization nor the location influenced the likelihood of participants correctly selecting the target string in the hierarchical menu.

6.4 Discussion

6.4.1 S3D for Change Detection and System Control

Our findings do not support the hypothesis that S3D has a statistically significant influence on task performance, driving performance, or simulator sickness during change detection or list selection tasks when using a large-scale S3D dashboard in a driving simulator. The location where the content is displayed showed also no statistically significant influence with the current design.

H1–H4 hypothesized that S3D influences secondary task performance and driving performance. In our study, we did not find a statistically significant influence of S3D. Looking at previous research, results are mixed. In some studies, e.g. Broy et al. [2014a], S3D led to improvements in secondary task performance. In others, no significant improvements (or deterioration) have been found (e.g. Broy et al. [2015a]). It is important to note that these studies used various combinations of visualization elements (e.g. color, form) whereas our study investigated the isolated effect of binocular disparity and the relative change of the elements size. Also, combining binocular disparity with others depth cues like occlusion, shadows, lighting, motion parallax (c.f. Cutting [1997]) might change results. For example, Broy et al. [2015a] showed that color positively has a stronger effect on perceived urgency when using S3D than with 2.5D.

The absence of any effect in our results can have several reasons. One of them is simulator sickness. *H6* postulated an influence of S3D on simulator sickness. There are no significant differences between groups, neither in the first, nor in the second task. Overall, the simulator sickness scores are rather high, reaching scores as high as 30.0 after the change detection task for the S3D group ($O_{max} = 159.18$, $D_{max} = 292.32$, $N_{max} = 200.34$, $T S_{max} = 179.52$). Especially in the change detection study, the elements on the dashboard showed high contrast. Here, the simulator sickness scores are rather high, even compared to the list selection task. The high contrast elements were introduced due to recommendations of Howard and Rogers [2008]. However, we believe that the many transitions from bright to dark in the change detection task might have led to elevated simulator sickness scores by introducing visual discomfort and visual fatigue. In addition to that, the vehicle physics—while based on NVIDIA physics—was perceived as unrealistic. In combination, these factors most likely led to the elevated simulator sickness scores. Hence, it is possible that the high scores mask any influence of S3D on simulator sickness. Also, it is possible that the overall high simulator sickness score negatively influenced task performance.

Second, the tasks did not really benefit from S3D besides offering a design alternative and presenting urgent or important information closer to the user. In our setup, S3D was mainly used to encode urgency or importance by displaying content closer to the user. Results indicate that urgency might not be enough to justify the application of S3D. Also, participants performed many iterations of the task and hence had a learning curve in an arguably rather easy task. It is likely that participants were simply too good in the tasks which makes S3D unnecessary. The low error rates support this theory.

While S3D as a plain design element resulted in better user experience or secondary task performance in previous research [Pitts et al., 2015; Szczerba and Hersberger, 2014], our results indicate that S3D does not necessarily work better but rather that it takes full effect only in certain situations and designs. Especially Pitts et al. [Pitts et al., 2015] and Szczerba et al. [Szczerba and Hersberger, 2014] used similar plain designs but their results indicate an increased secondary task performance. In other related research, no increase in secondary task performance is reported (c.f. Table 6.1)

6.4.2   Limitations & Future Work

Driving simulator studies have the disadvantage of only simulating the visuals and physics of real-world elements. We acknowledge this limitation. In this study, only negative parallax has been observed, meaning that highlighted objects appeared closer to the participant. This decision was made because prior work showed that displaying important content closer to the users is beneficial for performance [Alper et al., 2011] and expected by the user [Ventä-Olkkonen et al., 2013].

Table 6.1 summarizes and shows that previous results on S3D and its influence on secondary task performance regarding change detection or object detection vary widely. While for some tasks, mixed (or positive) results exist, it is still unclear

Table 6.1 Comparison with previous similar studies ($-$ = S3D worse, o = mixed results or equal, + = S3D better)

Author	Task type	Driving?	Color?	Better secondary task performance with S3D?
Krüger [2008]	Change detection on IC	y	y	$-$
Broy et al. [2015a]	Change detection on IC	y	y	$-$
Pitts et al. [2015]	Find depth difference	y	n	+
Szczerba and Hersberger [2014]	Change detection on IC	n	y	+
	Object detection on IC	n	y	o
Our results	Change detection on whole dashboard/subareas	y	n	o

what types of tasks and designs, binocular depth cues can support users. Interestingly, results of Pitts et al. [2015] were positive whereas our results were neutral. They performed a similar experiment (driving task, grey-scale graphics) but displaced the stimulus away from the user by increasing disparity. In our experiment, the torus moved towards the user (decreasing disparity). That indicates that moving an element away from the user is more effective than moving it towards the user. Our results add to the existing literature by providing evidence for the fact that an isolated application of S3D does not increase change detection or list selection performance while driving. It also opens up three research directions: first, considering that some of the previous related experiments have been performed without a driving task, what influence does a primary task have on the effectiveness of binocular depth cues? Second, what design elements can be used with binocular depth cues for enhancing peripheral detection (change detection) and list selection (object detection) performance? Color seems to have an influence as indicated by Broy et al. [2015a]. However, other elements like contrast, shape, or texture also influence binocular fusion and thus, might influence secondary task performance. Third, a more detailed classification on what kind of non-spatial tasks benefit from binocular depth cues and why they do it, is necessary to further build an understanding of binocular depth cues.

In summary, our results indicate that the application of S3D does not necessarily improve secondary task performance. Namely, (peripheral) change detection and list selection while driving. Contrary to some previous results, the isolated application did not measurably support drivers. Hence, using S3D plainly to communicate a change or to highlight an object—in a simple task and without spatial properties—seems to be inappropriate when the objective is to improve secondary task performance and by that, reduce driver distraction.

Using S3D for Trust and Social Presence 7

Parts of this section are based on Knutzen et al. [2019].

7.1   Problem Statement

Chapter 1 mentioned important problems of automated driving. Among them is the management of trust between the human and the automated vehicle. If humans do not trust the automation system (or any other ADAS), they might not use it at all or even use it in a wrong way [Lee and See, 2004]. The car, its behavior, and the user interface, have an influence on the human's trust in the automation system. Previous research tried to enhance trust by various methods, e.g. speech output [Koo et al., 2015] or a by making the system more transparent [Rezvani et al., 2016]. Another way to foster trust is to design the UI in such a way that it has human-like characteristics [Kulms and Kopp, 2019]. Such an attribution of human characteristics to an object is called *anthropomorphism*[1].

Previous research suggested that anthropomorphized user interfaces and agents can foster trust towards automated vehicles [Hock et al., 2016; Waytz et al., 2014]. Such agents can be applied in automated vehicles as assistants, representing the

---

[1] https://www.lexico.com/definition/anthropomorphism, Oxford English Dictionary, "Anthropomorphism", 2020-09-23

Electronic supplementary material   The online version of this chapter (https://doi.org/10.1007/978-3-658-35147-2_7) contains supplementary material, which is available to authorized users.

automation system. Then, they can explain the vehicles actions while it is driving automatically. By that, the agent creates a transparent and understandable environment for the users. They are then more likely to comprehend the vehicles actions. Previous research also showed that S3D can increase anthropomorphism (e.g. Kim et al. [2012]. In this case study, we take up the issue of trust by investigating whether and how an anthropomorphized agent in form of a humanoid representation displayed on the S3D dashboard can foster trust by explaining the automated vehicles decision.

7.2   On Trust in Virtual Anthropomorphized Agents

7.2.1   Trust

Trust is a multidimensional concept that describes the relationship between a trustor and a trustee. Many definitions have been derived over time. Considering the research area of trust in automation, Lee and See [2004] provide the definition that trust is "[...] the attitude that an agent will help achieve an individual's goal in a situation characterized by uncertainty and vulnerability". Two aspects of trust are behavioral trust (i.e. compliance) and self-reported trust [Muir, 1994]. The former describes the actions of a participant and how much they reflect a trustful relationship between him/her and the system. The latter describes the perceived level of trust a participant exhibits towards a system.

Trust needs to be calibrated, meaning that the amount of trust a human exhibits towards in a system needs to match the system's capabilities. Ill-calibrated trust can lead to inappropriate use or disuse [Parasuraman and Riley, 1997]. Reasons for this can either be overtrust or distrust. Overtrust means that the human's trust in the system exceeds its capabilities. Vice versa, distrust means that the human's trust in the system falls short of the system's capabilities, leading to disuse. A detailed explanation of trust in automation is provided by Lee and See [2004]. It is important to note that trust is a fragile concept, and that it takes time to build up a trustful relationship [Slovic, 1993]. Trust can be destroyed by several actions, e.g. unreliability, lack of authentic communication, or when breaking promises [Paliszkiewicz, 2017].

Overall, many factors influence trust between a human and a (semi-)autonomous system. For example, Hancock et al. [2011] list features that have shown to be important in human-robot interaction. Figure 7.1 shows the classification of these factors. It is important that not only the features of the system, but also those of the user (e.g. prior experience) and the environment (e.g. the type of task) need to be considered.

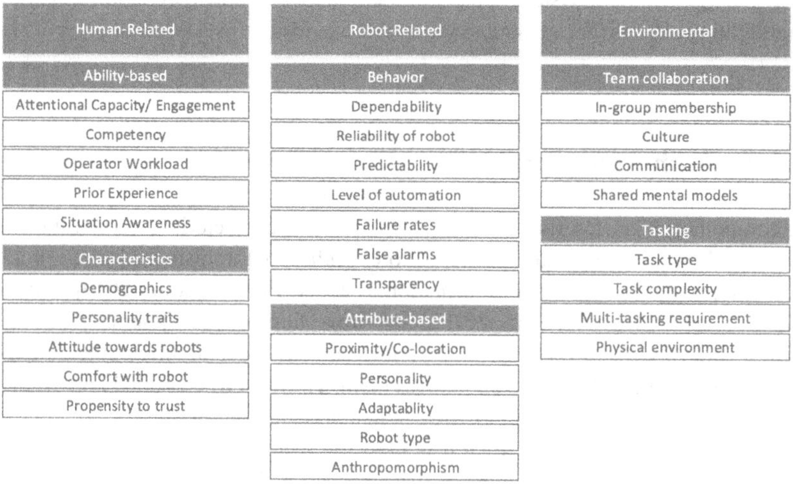

Human-Related	Robot-Related	Environmental
Ability-based	**Behavior**	**Team collaboration**
Attentional Capacity/ Engagement	Dependability	In-group membership
Competency	Reliability of robot	Culture
Operator Workload	Predictability	Communication
Prior Experience	Level of automation	Shared mental models
Situation Awareness	Failure rates	**Tasking**
Characteristics	False alarms	Task type
Demographics	Transparency	Task complexity
Personality traits	**Attribute-based**	Multi-tasking requirement
Attitude towards robots	Proximity/Co-location	Physical environment
Comfort with robot	Personality	
Propensity to trust	Adaptablity	
	Robot type	
	Anthropomorphism	

Figure 7.1 Factors influencing trust between humans and robots (adopted from Hancock et al. [2011])

In the automotive domain, trust towards the vehicle has been investigated in various experiments: Abe et al. [2018] tested three different driving styles that differed in lateral distance between an obstacle and the vehicle, the peak driving speed, and the first steering input when evading a foreign object on the road. Results indicate that the behavior of the vehicle, specifically the driving style, significantly impacts trust. Also, related to the vehicle behavior, Koo et al. [2015] conclude that to in order to build and maintain trust, vehicles should explain their actions. In their experiment, they did this by providing voice messages telling the driver-passenger why something happens and how the car is going to act. However they conclude that, while it is necessary to provide information to foster trust, too much or too little information can have adverse effects. Rezvani et al. [2016] investigated how a visual user interface that displays the awareness of the automation system, can foster trust. In their experiment, various dashboard visualizations were displayed to the participants, including confidence indicators and various levels of details. Their results indicate that information about the automation systems perception of a hazard had positive effects on trust (external awareness) whereas information about the current confidence level of the vehicle had negative effects (internal awareness). Similarly, Wintersberger et al. [2019] have evaluated AR displays, specifically an AR-HUD and its impact on trust in an automated vehicle. Researchers also conclude that communicating a systems upcoming decision [von Sawitzky et al., 2019] or highlighting objects that are hard to see by AR can foster trust [Wintersberger et al., 2017].

7.2.2   Anthropomorphism

Anthropomorphism describes the attribution of human characteristics to non-human entities[2]. Research and industry have explored anthropomorphism for various products, especially for intelligent information systems [Pfeuffer et al., 2019], robots [Schaal, 2019; Złotowski et al., 2015], and virtual humans [de Visser et al., 2016; Nowak and Biocca, 2003]. Objects or entities can have several different anthropomorphic properties. Table 7.1 summarizes them. They can be classified into visual, auditory, social, and mental properties. When applying and designing an anthropomorphic UI — especially when it is used for an investigation of trust — it is important to select manifestations of these properties in such a way, that trust is appropriately calibrated. For example, if the anthropomorphic UI should be used to perform critical task (e.g. evading an obstacle in an automated vehicle), it should show sufficient cognitive intelligence and that it has agency [Waytz et al., 2014].

In the automotive domain, the concept of anthropomorphism plays a role in several areas. People often see their car as a person [Aggarwal and McGill, 2007]. Also, the style of the car inspires implicit perception of certain traits and attributes

Table 7.1 Features influencing anthropomorphism

Category	Example of influencing property
Visual	Appearance, movements, gestures, mimics, gender, age, style (zoomorphic, anthropomorphic, caricatured, functional)
Auditory	Speech synthesizer, gender, age
Mental	Cognitive intelligence (dialogue ability, context, content understanding, image processing, speech recognition), emotionality, and emotional intelligence, personality, autonomy, imitation, moral values, accountability
Social	Social background, in-group/out-group

---

[2] See footnote 1

[Windhager et al., 2008]. Further, Gowda et al. [2014] have argued that an automated car explicitly should show character to increase user experience. Following up on these findings, Okamoto and Sano [2017] have previously proposed a prototype of an anthropomorphic AI agent including an abstract and "blob-like" representation to foster trust and increase UX. However, Qiu and Benbasat [2005] state that the absence of facial expressions can degrade the warmness and fidelity of the agent, making an anthropomorphic agent less efficient. Similarly, Broadbent et al. [2013] showed that robots with more human-like faces are attributed more mind and a better personality.

Prior research also points out that depth cues and especially binocular disparity can increase richness of interaction with media. Chuah et al. [2013] suggest that motion parallax is an important aspect during human-agent interaction, increasing the attractiveness of its appearance. Participants of Pölönen et al. [2009] state in a post-experiment interview that they felt that movie actors appeared more alive and their feelings towards them were more personal if they perceived footage in S3D. In more interactive settings, it has been shown that S3D can enhance anthropomorphism and overall communication in video conferencing [Hauber, 2012; Hauber et al., 2006; Muhlbach et al., 1995] games [Takatalo et al., 2011], and when interacting with virtual agents [Ahn et al., 2014; Kim et al., 2012].

Anthropomorphism is also related to trust. In a game, subjective ratings of the participants suggest that higher levels of anthropomorphism increase self-reported trust [Kulms and Kopp, 2019]. However, behavioral trust did not increase, meaning that participants did not act differently but perception of the virtual game partner changed. It has also been shown that anthropomorphism can increase trust in automated vehicles [Waytz et al., 2014]. In a driving simulator experiment, participants were required to drive in either a manual car, an automated car without an anthropomorphized agent, or an automated car with an anthropomorphized agent. Name, gender, and a voice were the anthropomorphic properties of the vehicle. The authors conclude that anthropomorphism improves trust, meaning that participants who drove the vehicle with anthropomorphic features reported higher self-reported trust. In another related study, Hock et al. [2016] let people drive an automated vehicle that was either equipped with no assistant, a voice assistant, or a virtual co-driver with voice. The latter was an augmented reality visualization visible via an AR-HMD and sat on the passenger seat. Their results indicate no difference in levels of trust. However, they report that the agent as a co-driver was perceived as uncanny and that the location was ill-chosen — participants had to move their head to see the co-driver: it was located on the passenger seat. They emphasize that an agent can be more useful if the sympathy towards it is large enough.

Despite the mentioned positive aspects of anthropomorphism in general and towards trust, it also has certain disadvantages that need to be considered. First, it is hard to design an anthropomorphic agent that is universally appealing [Culley and Madhavan, 2013]. For example, skin color can lead to in-group bias which, in return, can negatively affect trust towards the agent [Eyssel and Loughnan, 2013]. Another problem that exists, especially with human-like anthropomorphized user interfaces, is the *uncanny valley*. Introduced by Mori et al. [2012] in 1970, the uncanny valley describes the relationship between humanlikeness of an entity and a person's affinity towards this entity. Originally designed for robots, it has also been applied to virtual humans (e.g. Seyama and Nagayama [2007]). The uncanny valley states that the affinity towards an object increases with increasing humanlikeness but steeply decreases for some interval when a certain level of humanlikeness is reached. This happens when the entity is already quite similar to a human. After this decrease in affinity, it increases again to higher levels than before. Figure 7.2 illustrates this concept.

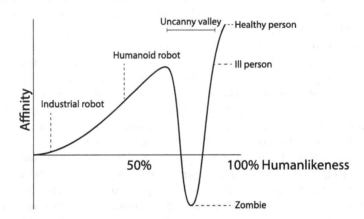

Figure 7.2 The uncanny valley. (adopted from Mori et al. [2012])

With that, previous research showed that trust is important when using or applying automated driving to prevent misuse or disuse. It has also been shown that anthropomorphized user interfaces can increase trust in automation. Further, there is evidence that depth cues and binocular depth cues can increase anthropomorphism, trust, and the richness of user interfaces. We take that as a motivation to investigate if a human-like agent that represents the automation system can increase trust during driving an SAE level 3 automated vehicle.

7.3   Study Design

When humans are driver-passengers in an automated vehicle, they have to rely on the vehicle. Hence, it is important to have an automation system, that the human sufficiently trusts. Related work shows that trust can be increased by considering factors affecting trust when designing a user interface. For example, by adding anthropomorphic cues to the user interface. This case study was designed based on findings from prior work, especially from Waytz et al. [2014] and Hock et al. [2016]. They showed that in automated vehicles, an anthropomorphized UI can increase trust and that a virtual agent, that is not likable and not visible, does not increase trust. We take up this research and investigate if an S3D visualization of a virtual agent can increase trust in the automation system of a vehicle. Based on the related work, the following hypothesis guide this case study:

H1:   An anthropomorphized agent, displayed in S3D, induces more behavioral trust than an anthropomorphized agent displayed in 2.5D.

H2:   An anthropomorphized agent, displayed in S3D, induces more self-reported trust than an anthropomorphized agent displayed in 2.5D.

To see if any influence of S3D depends on the level of trust, we apply two levels of anthropomorphism. While we do not change the visual representations of the agent, we remove other anthropomorphic cues. We further collect data on anthropomorphism to verify the agent's design.

This case study employs a 2 x 2 between-subject design. The first between-subject variable is the visualization condition with two levels: 2.5D or S3D. The second between-subject variable is the level of anthropomorphism with *less* and *more*. Overall, the procedure is inspired by the previous study of Hock et al. [2016] who investigated the influence of anthropomorphism on trust in automated driving.

7.3.1   Procedure

When participants arrived in the laboratory, they got an explanation of the study and its purpose. They then signed a consent form (informed consent). They had to pass a stereo vision test [Schwartz and Krantz, 2018]. Then, participants were assigned to one of four groups (combinations of 2.5D and S3D as well as less anthropomorphism and more anthropomorphism) following a counter-balanced procedure. After that, they took a seat in the simulator. They got an explanation of the simulator and took a test drive to get familiar with controls. Then, the experiment started. Participants began driving and had to enable automation. With that, they were greeted by the agent. The car drove along a two-laned road with approaching traffic. Mean speed was 100 km/h at this time. After 15 seconds, two vehicles appeared on the opposite lane, emphasizing the possibility of approaching traffic. After another 60 seconds, the car encountered a lorry on its lane and slowed down to drive behind him with a speed of 60 km/h. It did not overtake because fog had obstructed vision. The agent provided this reason on why it is not able to overtake. After 45 seconds, the agent again told the driver-passenger that it cannot overtake due to restricted vision. Now, every 40 seconds, the visual range, which was initially 200 m, increased to 500 m, then to 800 m, and finally to 1100 m. These values were adopted from Hock et al. [2016]. With every change, the agent tells the user that it cannot overtake due to limited vision. This is supposed to create a conflict in driver-passengers because, at some point, they might think that it is safe to overtake. At this point, the participants might take over control and by that, not trust the judgment of the automation system. If they took over control, the agent acknowledged that by issuing a statement. After the last change to 1100 m visual range, the vehicle continues driving for another 45 s. Then, the agent ended the drive by issuing a goodbye message. The participants then filled out the post-experiment questionnaires. With that, the experiment ended. Participants were not paid.

7.3.2   Apparatus

Agent
The design of the agent was carefully selected considering the mentioned features influencing anthropomorphism and trust. The 3D model was created using Blender 2.79 and animated using the extension MB-Lab[3]. Our agent featured a female human-like face including hair. It was decided to use a humanlike model because previous research showed an increased level of trust in such models [Verberne, 2015; Yee et al., 2007]. The model showed a medium level of realism to avoid the uncanny valley but to still appear human [Mori et al., 2012]. The agent is supposed to have a medium level of attractiveness [Mori et al., 2012]. To make the agents appearance less perfect, it had freckles, redness, decent make-up, an uneven skin color, and beauty spots [Mori et al., 2012]. Because we mostly expect to have participants from central Europe, a Caucasian hair-, skin-, and eye-style was implemented [Verberne, 2015].

---

[3] https://github.com/animate1978/MB-Lab, MB-Lab, 2020-09-23, previously available at http://www.manuelbastioni.com/

Facial expressions were added as an additional anthropomorphic feature to further enhance humanlikeness by enabling facial movements [de Melo et al., 2014;
Lee and See 2004; Nass and Moon, 2000]. We implemented a neutral face, a smirk-
ing face, and a smiling one. Various views of the agent are shown in Figure 7.3.
If the agent talked to the user, it looked towards the user [Hoff and Bashir, 2015].
To do this, the tracking target of the stereo glasses was used to get the position and
orientation of the user's head. The agent's eyes blinked in pseudo random intervals
between 2 and 5 seconds.

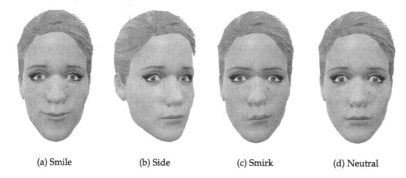

    (a) Smile   (b) Side   (c) Smirk   (d) Neutral

Figure 7.3 Different expressions and views of the agent

The agent's audio sentences were recorded with a female non-professional voice
actor. Research indicates that female voices are more trustful in recommender systems [Payne et al., 2013; Wintersberger et al., 2016]. For recording, a Technica
AT 4035 SV large-diaphragm condenser microphone (frequency response: 20 Hz–
20.000 Hz, open circuit sensitivity: 0 dB), a RME Fireface 802 decoder module,
and Logic ProX were used. Recordings were done with a sample rate of 44.100 Hz
and 32 bit (lossless format). The 3D model's motions were matched to the sound
phonemes of the recordings using Papagayo[4]. By that, the agents mouth moved
according to the spoken sentences. The implemented sentences are listed in Table
7.2. The introduction as well as goodbye statements were added and phrased so
that they add purpose and fulfill social expectations. They should transport a polite
image of the agent and communicate name and gender. By adding these details and
following a social protocol, the agent is supposed to appear more like a social actor
and by that more anthropomorphic.

---

[4] https://my.smithmicro.com/papagayo.html, Papagayo, 2020-09-24

The agent had two behavioral states: *idle* and *communicate*. In idle mode, it looked around using head and eyes, occasionally smiled, and blinked from time to time. This mode was added to simulate human behavior and by that, increase anthropomorphism [Verberne, 2015]. The head could rotate in 3 DOF (roll, pitch, and yaw). The eyes could rotate in 2 DOF (pitch and yaw). In communication mode,

Table 7.2 Verbal sequences of agent including wording and trust-affecting and anthropomorphic cues

Event	Wording	Trust-affecting and Anthropomorphic Cues and Reference
Introduction	"Hello, it's nice to meet you. My name is Mary. I'm your driving assistant. I'm in charge of the vehicles speed and steering. I inform you when your intervention is required in relevant situations. You can take over control from me at any time if you'd like to. I'm looking forward to work with you."	Purpose and social expectation [de Visser et al., 2012], politeness & etiquette [Parasuraman and Miller, 2004], gender [Waytz et al., 2014], social actor [McCallum and McOwan, 2014], name [Laham et al., 2012]
Deceleration	"I sense a car in front that we're approaching too fast. Therefore, I adapt the vehicle speed to 100 km/h."	
Cannot overtake	"The sight is restricted. I'll overtake as soon as possible." "I can't see clearly. Therefore I can't overtake the car ahead." "The fog is very dense. Therefore I can't overtake the car ahead." "I can't see approaching traffic. There- fore I can't overtake the car ahead." "The fog is still too dense. Therefore I can't overtake the car ahead."	What and why [Hock et al., 2016], [Kraus et al., 2016]
Control transfer	"You switched to manual mode. Therefore, I abort automation functions."	
Saying Goodbye	"The drive ended. I hope you enjoyed the trip. Have a nice day. Goodbye."	Purpose & social expectation de Visser et al. [2012], politeness & etiquette [Parasuraman and Miller, 2004]

the agent looked towards the user and talked using the implemented sentences and lip sync. The behavior of the agent was controlled by the driving simulation. At predetermined intervals, the agent reacted with the sentences that state reasons for non-overtaking. At the beginning and the end, greeting and goodbye are triggered. While not talking, the agent is in idle mode. For creating and maintaining eye contact, we use the position of the active 3D glasses. Disparity of the agent's farthest point was D_{max} = 0.66° and of the closest point D_{min} = −0.61° . The agent was displayed on the right upper side of the dashboard all the time as illustrated by Figure 7.4b. This location was chosen because participants showed a good change detection behavior in this area (c.f. Section 6.3.2 but also because it offered a relatively large zone of comfort (c.f. Figure 5.2c).

We further included two levels of anthropomorphism: *less* and *more*. In the *less*-condition, participants only heard the messages listed under deceleration, cannot overtake and control transfer. These participants did hear have any other of the trust-affecting cues. In the *more*-condition, the agent introduced herself and provided information about what she is able to do. She also provided a goodbye message to the user at the end of the drive. By that, we add purpose, social expectation, politeness, etiquette, gender, and a name as trust-affecting cues and hope to make the agent a more valid social actor.

Driving simulator

The driving simulator had the design as described in section 4 including head-tracking for off-axis projection and coated boarders.

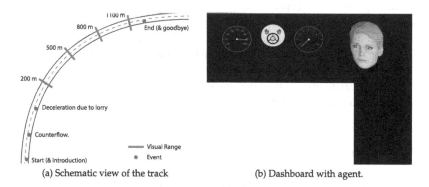

(a) Schematic view of the track   (b) Dashboard with agent.

Figure 7.4 Technical details of the S3D avatar experiment: Figure 7.4a shows the course employed during the experiment. Figure 7.4b shows the agent on the dashboard

The dashboard showed the agent, an automation indicator, and two gauges — speed and rpm. The automation indicator turned green when automation was enabled, red when enabling automation was permitted and grey when automation was off but could be enabled. Automation can be disabled by turning the steering wheel more than 5° towards either side or by hitting a pedal more than 10% [Gold et al., 2013]. Initially, we intended to use a wheel turn angle of 2° rule instead of 5° . However, backlash (or play) and inaccuracies of the wheel forced us to increase the value to 5° . In total, moving the steering wheel 5° modifies the direction of the vehicle by 2° (the wheel has 3° play). In this SAE level 3 automation system, the car was able to hold the lane, accelerate, and decelerate. Automation could also be disabled by another dedicated button on the steering wheel. The driver needed to be fallback ready all the time.

A schematic view of the track is illustrated in Figure 7.4a with the viewing distances and other relevant events. Driving simulation was built with Unreal Engine 4.18, the dashboard visualization with Unreal Engine 4.23.

7.3.3   Sample

56 participants took part in the study (80.9% male, 20.1% female). The youngest was 19 years and the oldest 52 years old (M = 24.78, SD = 4.89). All had normal or corrected to normal vision, a valid driver's license from their country, and passed a stereo vision test [Schwartz and Krantz, 2018]. All of them had normal or corrected to normal vision. Participants were invited using convenience sampling by mailing lists, direct contact, and Facebook groups. The final sample consisted mostly of students and staff of Technical University Ilmenau. 88.2% had previously experienced S3D displays, e.g. in a cinema. 42.5% previously drove in a driving simulator. 66.2% have a car and 37.8% of those regularly use an ADAS (e.g. active cruise-control or lane-keeping assist system). There were no dropouts. Mean motion sickness susceptibility score of the sample M = 8.387 (SD = 9.40). Average experiment duration was 35 minutes.

7.3.4   Measures

To measure trust, participants had to fill out the *System Trust Scale* questionnaire developed by Jian et al. [2000] for self-reported trust. We further measured behavioral trust via the time it took participants to take over control from the automation

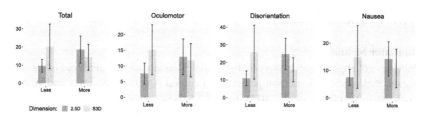

Figure 7.5 Mean values and standard deviations of the simulator sickness symptoms (lower is better; O = Oculomotor [0;159.18], N = Nausea [0;295.74], D = Disorientation [0;292.32], TS = Total Score [0;235.62])

system in accordance to Hock et al. [2016]. We also measured the proportion of people who passed another vehicle without having appropriate vision to those who did so with appropriate (or safe) range of vision. A passing maneuver was considered safe if it has been performed in a segment where the range of vision is higher than 500 m. The German Federal Highway Research Institute recommends a safe range of vision of 600 m on rural highways with regular non-slowed traffic [Bundesanstalt für Straßenwesen, 2017]. In our experiment, traffic was slowed down to 60 km/h. Thus, we decreased the value slightly to classify a safe distance. We measured user experience via the user experience questionnaire (UEQ, Schrepp et al. [2017]) and anthropomorphism using the questions introduced by Waytz et al. [2014]. To further assess the agent's design, we added own questions to get insights on the participant's view of the agent. Participants further had the option to provide written feedback on the agent in a text field. Questionnaires can be found in the electronic supplementary material C2.

7.4 Results

Data was analyzed using R 3.6.1 (bestNormalize_1.4.0, car_3.0-3, and ARTool_0.10.6). Normal distribution of the residuals was assessed by Shapiro-Wilk tests and QQ-Plots. Significance criterion was $\alpha = 0.05$. Outliers were removed for values above or below 2 times the inter-quartile range [Elliott and Woodward, 2007]. When ANOVAs were applied, homoscedasticity was given if not stated otherwise. Pairwise comparisons were performed using Tukey HSD tests. 56 participants were equally divided among the four groups (14 per group). There were no dropouts.

7.4.1   Control Variables

Simulator Sickness

Data was not approximately normal distributed. Hence, data was analyzed using an aligned rank transform and non-parametric factorial ANOVAs [Stanney and Wobbrock 2019; Wobbrock et al., 2011]. Figure 7.5 shows the results of the SSQ data. Results did not detect statistically significant main or interaction effects in simulator sickness ($F(1, 52) < 1.95, p > .168, \eta_p^2 < 0.036$). That means that neither the level of anthropomorphism nor the visualization technique led to different experiences in simulator sickness. The electronic supplementary material C.1 shows results of the statistical analysis.

Anthropomorphism

Figure 7.6 shows results of the anthropomorphism questionnaire. Data of the anthropomorphism questionnaire was approximately normal distributed after applying a Yeo-Johnson transformation [Yeo and Johnson, 2000]. Results of a two-factor ANOVA indicate a significant main effect of visualization on anthropomorphism in the Yeo-Johnson-transformed data ($F(1, 52) = 5.78, p = .020, \eta_p^2 = .10$). That tells us that displaying the agent in S3D leads to significantly higher levels in anthropomorphism than a 2.5D visualization. The absence of a main effect of *Level of Anthropomorphism (Level-A)* tells us that the intended different levels of anthropomorphism (less and more) did not lead to different perception of anthro-

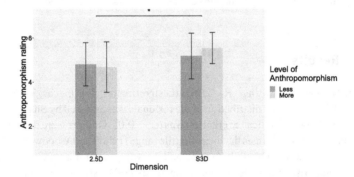

Figure 7.6 Untransformed results of the anthropomorphism questionnaire. There was a significant difference between 2.5D and S3D. (min = 1, max = 7, higher values indicate more perceived anthropomorphism, significance codes: *** : p < 0.001,** : p < 0.01,* : p < 0.05)

pomorphism. There were no other main or interaction effects. Full statistical data can be found in the electronic supplementary material C3.

Attitude towards the Virtual Assistant
Attitude towards the assistant was assessed using five statements on a Likert scale from 1 to 5. Participants had to rate if they strongly disagree, disagree, neither disagree nor agree, agree, or strongly agree with them. The statements are as follows (information in brackets was not shown to participants but added here to support interpretation):

1. The assistant creeped me out (lower better).
2. The assistant's appearance matched its abilities (higher better).
3. I found the assistant's notifications annoying (lower better).
4. The assistant distracted me from the driving task (lower better).
5. I felt uncomfortable when the assistant looked at me (lower better).
6. It is all right that a driving assistant takes a high level of responsibility (higher better).

Figure 7.7 Untransformed results of item asking if glances were uncomfortable. There was a significant difference between conditions *Less* and *More* (1 = strongly disagree, 5 = strongly agree, lower values indicate lower uncomfortable feelings, *** : p < 0.001,** : p < 0.01,* : p < 0.05)

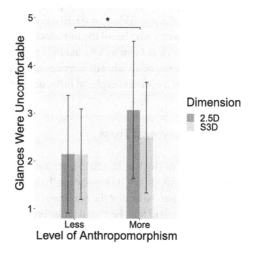

Table 7.3 Mean and standard deviation for the single items describing participants attitude towards the virtual assistant (1 = strongly disagree, 5 = strongly agree)

	M	SD
Is creepy	2.554	1.127
Ability matches appearance	3.339	0.920
Is annoying	2.946	1.182
Is distracting	2.286	1.004
Glances were uncomfortable	2.464	1.235
Agree with responsibility	3.589	0.968

Table 7.3 shows the overall mean values and standard deviations of participants' ratings. The electronic supplementary material C.4 shows the detailed and group-wise descriptive statistics. Data of all scales was not approximately normal distributed. Hence, a series of univariate ANOVAs on aligned rank-transformed data was performed. The results of these ANOVAs indicate only one main effect. Participants who had experienced the avatar that was classified as low anthropomorphism rated the glances of the agent less uncomfortable (M $=$ 2.14, SD $=$ 1.08) than those who experienced a high-anthropomorphism-version (M = 2.79, SD = 1.32) ($F(1, 52) = 4.669, p = .035, \eta_p^2 = 0.010$). Figure 7.7 illustrates this difference. It tells us that participants who heard the introduction and the goodbye message perceived the agent looking at them as less annoying than those who didn't hear these messages. The absence of all other differences indicates that neither type of visualization nor the level of anthropomorphism influenced other items.

7.4.2 Behavioral Trust

In accordance to Hock et al. [2016], we assume that the later participants waited to take over control, if they took over at all, the more they trusted the system to make the right decision. Note that the automation system always decided to not overtake and, as a reason said that the visual range is not good enough to overtake. For people who did not overtake, we noted down 240 seconds as the control transfer time (the maximum duration of one drive).

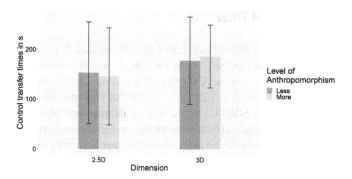

Figure 7.8 Descriptive statistics of the time it took participants to take over control in seconds (untransformed, min = 0, max = 240 seconds, higher values indicate more behavioral trust)

Figure 7.8 illustrates our data on control transfer. Data on control transfer times was not approximately normal distributed. A Shapiro-Wilk test revealed a highly negatively skewed data distribution (W = 0.795, p-value < 0.001, skewness: x = − 0.491). Common data transformations (Box-Cox, Log, square root, exponential, Yeo-Johnson, order-norm, or inverse hyperbolic sine) did not lead to normal distribution. Results of univariate ANOVAs on aligned-rank-transformed data [Wobbrock et al., 2011] indicate no significant main or interaction effect in control transfer times $(F(1,52) < 1.802, p > 0.185, \eta^2 < 0.033)$. That indicates that neither the type of visualization nor the implemented differences in anthropomorphism influenced the level of behavioral trust measured by the control transfer time. Complete statistical data is listed in the electronic supplementary material C.5.

To analyze the ratio between safe take-overs and unsafe take-overs, Fisher's exact Test was used. Results did not indicate significant differences between proportions in the 2.5D and S3D condition (p-value = 0.143, CI = [0.704;45.89], odds ratio = 4.226) That tells us that the proportion of safe take-overs is not higher in the S3D condition than in the 2.5D condition and by that, people experiencing the S3D condition did not perform statistically more safe take-overs than those experiencing the 2.5D agent. Unfortunately, from the original research of Hock et al. [2016], no count data for a contingency table was available. Hence, no Fisher's Exact Test (or another proportion test) could be calculated as a comparison.

7.4.3 Self-reported Trust

Figure 7.9 presents the descriptive statistics of the self-reported Trust questionnaire (System Trust Scale). Data of the System Trust Scale was approximately normal distributed. Results of a two-factor ANOVA indicate no significant main or interaction effects ($F(1,52) < 1.51$, $p > 0.223$, $\eta^2 < 0.028$). That indicates that neither the type of visualization (2.5D/S3D) nor the level of anthropomorphism (less/more) had an influence on self-reported trust between human and automation system. The results of the ANOVA can be found in the electronic supplementary material C.6.

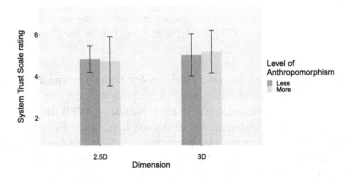

Figure 7.9 Results of the System Trust Scale. There were no significant differences (min = 1, max = 7, higher values indicate more self-reported trust)

7.5 Discussion

We detected no differences in simulator sickness between groups. Hence, we can assume that different levels of simulator sickness did not influence results. Overall scores range between 7.58 for nausea in the 2.5D visualization group who did not experience any introduction or goodbye message to 25.85 for disorientation in the 3D group that did not experience an introduction or goodbye message. While these scores can be considered as elevated, they are in line with scores provided by Stanney et al. [2010] who evaluated eight different virtual environment simulators.

7.5.1 Evaluation of the agent

The design of the agent achieved a reasonable level of anthropomorphism. The overall mean value $M = 5.049$ (SD = 1.011) on a scale from 1 to 7 (7 is the best) indicates that participant perceived the agent who represented the automation system as a humanlike entity.

However, the intended two levels of anthropomorphism did not influence the reported level of perceived anthropomorphism. The two conditions differed only by having or not having a greeting and goodbye message. It is likely that these messages are too short to make a difference. In addition to that, it is possible that the behavior of the assistant and the visuals (which were present the whole drive), had a much higher influence than the short audio messages.

Interestingly, S3D had an effect on reported anthropomorphism. The final score was significantly different in the S3D condition ($M_{S3D} = 5.357$, SD = 0.891), compared to the 2.5D condition ($M_{2.5D} = 4.741$, SD = 1.044; min = 1, max = 7). Hence, it can be assumed that the spatial presentation using binocular depth cues and motion parallax has a positive influence on perceived anthropomorphism. This is in line with previous results (c.f. Section 7.2.2 where S3D and/or motion parallax increased anthropomorphism. Hence, we can assume that the two presentation modes were perceived differently regarding anthropomorphism.

The measures assessing the participants' attitude towards the agent can be considered neutral or slightly positive. Answers indicate that the agent was not perceived creepy, albeit it also was not perceived as especially pleasant. Participants were slightly in favor with the statement that the agent's ability matches its appearance. They also responded that the agent was not annoying, but they also didn't think that it was especially enjoyable. The agent was not considered distracting. Participants did also not agree with the statement that the glances of the agent are uncomfortable. All participants also agreed more with the driving assistant taking over a high level of responsibility.

Summarizing, the results of this section indicate that the participants had a rather neutral attitude of the agent. All mean values indicate a slight positive attitude and participants' attitude towards the agent can be described as reserved. By that, we can assume that the agent's design was chosen appropriately so that it does not causes excessively negative feelings which might negatively influence results. Yet, a more positive attitude might also influence trust, considering that attitude is a factor influencing trust. Hence, future agents should focus on a more appealing design.

The qualitative feedback that some participants provided, gives some indications and a first explanation on the rather neutral attitude of participants. On the one hand, some participants were critical towards the agent. They mentioned that a face is not necessary and voice would be enough (P63: "It's not necessary to show [the] face of

[the] assistant".), that the voice should not give updates on what it does but just when it is doing something (P44: "I also don't need frequent updates when the condition that prevents it from taking certain actions doesn't change.", and that the feedback sounded repetitive (P59: "Less repetitions, not telling so much"). Note that the repetitions were intentional to instill distrust. On the other hand, some participants also appreciated the design (P48: "The face looks very like a human and is very cool.", P51: "I loved the details in her skin and that she followed me with her eyes.") but also mentioned that they would have liked customization options (P15: "I believe that [customizing it] will allow to relate and believe in the assistant a lot easier.", P61: "To be able to choose the voice or 3D picture of driving assistant."). Some also mentioned that — while the agent was fulfilling its purpose — they would have liked to explore its capabilities more, for example, with voice commands (P51: "Would have loved to see what else she could do").

Overall, the attitude of the participant can be considered neutral or slightly positive. This lets us assume that the evaluation of trust measures in the next section is based on an agent that was not perceived negatively but received balanced opinions.

7.5.2   The agent and trust

We expected that the visualization of an agent using binocular depth cues (in S3D) will increase behavioral and self-reported trust. In other words, we expected that participants will pass the leading vehicle later or not at all (increased control transfer time, H1) and that they will rate the agent more trustworthy via the System Trust Scale questionnaire (H2).

Overall, participants trusted the agent rather well. Results of the System Trust Scale questionnaire are slightly positive with a total mean of $M = 4.927$ (SD = 0.972). Hence, we can assume that the agent was perceived as a rather trustworthy entity. Results assessing behavioral trust back these results. Of all participants, more than half (n = 28) did not initiate a passing maneuver at all, indicating that they trusted in the agent's decision to not pass. 21 more issued a control transfer only when having a safe visual range. While those participants didn't trust the assessment of the agent, this distrust at later stages did not lead to misuse. Only 9 participants terminated automated driving while visual range was not sufficient. However, results of our study indicate no statistically significant effect of the independent variables. We did not detect any significant difference of the type of visualization (2.5D or S3D) on self-reported (H1) or behavioral trust (H2).

The high standard deviations of the measured control transfer times indicate that there are important aspects that our study design did not take into account. For example, Hegner et al. [2019] mentioned that, besides trust, driving enjoyment, personal innovativeness, and perceived usefulness influence the acceptance of autonomous vehicles (e.g. some participants stated that the agent was not necessary during the final feedback session.) Further, while the agent was perceived neutral by the participants, its design arguably can be further improved.

It is interesting to note that participants considered the S3D agent to be more anthropomorphic than the 2.5D agent. Previous research indicates that higher levels of anthropomorphism lead to more trust. In our study, this was not the case. One possible explanation is that the depth cues did not increase anthropomorphism enough to manifest a measurable effect. Also, previous research incorporated more interaction with the agents (e.g. direct communication [Ahn et al., 2014; Gotsch et al., 2018; Kim et al., 2012]). In our experiment, interaction with the agent was limited. To explore this further, we calculated the correlation between anthropomorphism and trust in our experiment (post-hoc). Results of a Spearman correlation indicate that there is a significant relationship between self-reported trust and anthropomorphism ($\rho_S = .39, p < 0.01$) but not between behavioral trust and anthropomorphism ($\rho_S = .05, p > 0.05$). That means that a higher score in self-reported trust came with a higher score in perceived anthropomorphism which is similar to previous results acquired in 2.5D video games [Kulms and Kopp, 2019]. That motivates further exploration of this correlation and the potential causation.

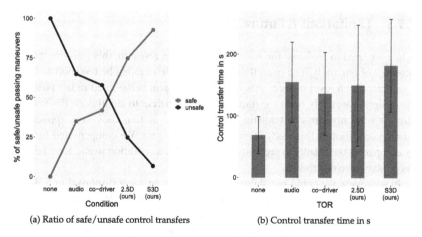

(a) Ratio of safe/unsafe control transfers   (b) Control transfer time in s

Figure 7.10 Comparison of our results on safe/unsafe maneuvers and control transfer times with previous results of Hock et al. [2016] (none, audio, and co-driver)

The high number of people who did not take over at all needs to be taken into consideration. In the evaluation, we included them with the maximum control transfer time of CT = 240 s. This group accounted for 28 participants or half of the total sample of N = 56. While we could assume that this group trusted the agent completely, it is also possible that they simply did not dare to take over control due to the laboratory setting, the stressful situation [Sonderegger and Sauer, 2009], or because they were not sure if they were allowed to (note, they were told that they could take control of the vehicle any time). Also, there was no inherent motivation for participants to initiate a passing maneuver, hence some participants might have chosen to simply wait and see what happens and if there would be some kind of instruction.

Nevertheless, comparing our results with the best condition in Hock et al. [2016] indicates an improvement of our agent compared to their virtual co-driver in terms of the number of safe and unsafe control transfers. Figure 7.10 illustrates the relation between our results and previous results. In our setting, participants performed more safe takeovers in total. Also average control transfer times are higher than in previous research. While it was not possible to calculate statistical tests to confirm significant differences due to absent data of previous research, data at least indicates positive effects of a virtual agent displayed on the dashboard.

7.5.3   Limitations & Future Work

While we tried to replicate the scenario of previous research, this also limited the potential of our study. The overall duration of the drive might be too short to build a sufficient high level of trust. Also, despite the agent being rated rather positive, the design is arguably open for improvements. Inherent to simulators, the vehicle physics was only close to realistic. In addition, no motion cues were applied. A missing condition in this study is one without any agent. We compensated for that by comparing our results to previous ones, but such a condition would have helped to contextualize our results.

In summary, we could not show an influence of an agent displayed in S3D on trust in automation while driving a semi-autonomous vehicle. While the agent was perceived as rather trustworthy and the attitude towards it was also above average, self-reported trust via the System Trust Scale and behavioral trust measured via control transfer times and the number of safe take overs did not differ significantly. However, we detected a difference in anthropomorphism. The S3D group rated the agent to be more anthropomorphic compared to the 2.5D group — a fact worth further investigations. Especially considering the positive correlation between anthropomorphism and self-reported trust. Potential modifications that further enhance anthropomorphism (e.g. a personalized agent or different designs) as well as situations that have a longer and dedicated trust-building phase are promising future research opportunities.

Urgent Cues While Driving: S3D Take-over Requests

8

Parts of this section are based on Weidner and Broll [2019c] and Weidner and Broll [2020].

8.1 Problem Statement

In a vehicle equipped with a SAE level 3 automation system, the human can hand over control to the vehicle [SAE International, 2015]. When this mode is activated, it executes the lateral and longitudinal control of the vehicle without the necessity of human intervention. However, the human still needs to take over control in case the vehicle encounters a situation that is outside its defined operational design domain. If it encounters such situations, it issues a take-over request (TOR) and the human has to take over control — at best immediately — and react safely.

Such a control transfer from the automation system to the human is called a take-over. A take-over can be user-initiated or system-initiated (in the previous section, Chapter 7, a human-initiated take-over was performed before the passing maneuver). A system-initiated take-over can be time-critical or scheduled. Scheduled describes a take-over that is known a comfortable time frame before it happens — e.g. when the system knows that it is going to leave its operational design domain on purpose (like a highway exit). A time-critical system-initiated take-over happens if the vehicle encounters a situation where it is not sure, not able, or not allowed to handle it, and

Electronic supplementary material The online version of this chapter (https://doi.org/10.1007/978-3-658-35147-2_8) contains supplementary material, which is available to authorized users.

did not expect this situation. For example, a complicated traffic scenario where a leading vehicle suddenly breaks. More information on take-overs can be found in McCall et al. [2016].

It has been shown that humans are rather bad when performing supervision tasks on automated systems. After a time, they get out-of-the-loop, meaning that their mind disengages from the supervision task and focuses on other tasks [Endsley and Kiris, 1995]. By that, they lose situation awareness. However, before taking over control, they need to have a certain level of situation awareness [Gold et al. 2015]. If they do not possess the necessary situation awareness, gaining it is necessary. Without this, it is hardly possible to react in a safe and appropriate way [Endsley 1995]. While motor readiness is quickly available [Zeeb et al. 2016], gaining situation awareness takes a certain amount of time. However, time is a crucial factor in these time-critical events. The higher the delay during a take-over, the higher is the risk of an accident Gold et al. [2013].

Various systems have been developed, trying to support the driver during these situations. In this case study, we investigate if binocular depth can succeed in supporting drivers during a system-initiated, time-critical take-over. In detail, we want to investigate if binocular depth cues can help to communicate spatial information faster and better than presentation on traditional 2.5D displays. Note that, compared to the study on driver distraction in Chapter 6, this study focuses on spatial information rather than using S3D as a plain design element for generic content. By that, we intend to improve gaining situation awareness of the current traffic situation.

Providing visual information might seem counter-intuitive at first because driving is a highly visual task [Wickens, 2002] and by that, a conflict over cognitive resources might happen. Hence, the added visual information might hamper take-over performance by actually distracting the driver rather than supporting him. However, such cues still might be necessary as previous research indicates that drivers either take-over without having assessed the situation at all [Walch et al. 2015] or that they do a complete visual scan of the environment when provided without any cues [Telpaz et al. 2015]. Here, providing curated pieces of information could make scanning the surroundings superfluous while helping drivers to get necessary information [Bueno et al. 2016]. Nevertheless, it is necessary to carefully control for mental load and visual information processing when adding visual information.

In our case study, we imagine a scenario where cars are connected and exchange data (e.g. type, color, but also positional data [Qiu et al. 2018]). While regulation or control algorithms of cars might not be ready to execute complicated evasive maneuvers, these data could be used to support drivers during take-overs.

In this case study, we explore S3D take-over notifications that provide basic information about the current traffic situation and hints about the danger that the vehicle sensed. Further, we investigate if presenting this spatial information efits from binocular depth cues. These two aspects are motivated by positive outcomes of prior research on directional take-over requests and stereoscopic information visualization. In addition to that, we assess if we can exploit the large dashboard by displaying the information in three ways: first, at the traditional location at the instrument cluster. Second, at the location where the driver is currently looking at (focus of attention). And third, by only displaying a large take-over request and hiding other content of the dashboard. Visualizations of take-over have been researched at the instrument cluster and center console but not on the focus of attention and also not by isolating the visualization from other visual content. By analyzing gaze behavior we further want to assess if the added visual information (in all three visualization scenarios) increases workload of the driver or negatively influences visual information processing. This is necessary as the additional visual information of a visual directional take-over request might led to a resource conflict during information processing. In summary, we hope to use S3D visualizations of the surroundings presented on the dashboard during a take-over maneuver to increase safety without impairing gaze behavior or increasing mental workload.

8.2   On Mental Workload and Take-over Notifications

8.2.1   Mental Workload While Driving

In a 2019 review, Butmee et al. [2019] state that mental workload has no clear definition and that it is a multidimensional construct. Other authors agree and add that it describes the relationship between the human's processing capacity, the task's total mental demand, and the context or environmental influences where the task is performed [Charles and Nixon, 2019; Marquart et al., 2015; Recarte et al., 2008]. In addition to that, the theory of multiple resources [Wickens, 2002] states that, if two processes need the same type of resources (e.g. visual), a conflict arises and a competition for the finite amount of visual processing capacity arises. In the case of driving and take-over, it has been argued that additional visual information in the take-over notification during the time-critical situation of a take-over can impair performance by creating such a conflict for visual processing capacity [Borojeni et al. 2018]. For us, that means that, it is possible that the performance decreases due to the increased mental demand of the task compared to a take-over notification without added information.

Mental workload can be measured in various ways. In the driving context, workload is measured by questionnaires (DALI [Pauzie, 2008], NASA TLX [Hart and Staveland, 1988], SWAT [Reid and Nygren, 1988]), performance-based measures, and physiological measures including ocular measures [Butmee et al., 2019]. For the traditional and often applied questionnaires NASA TLX (NASA Task Load Index) and the derived DALI (Driver Activity Load Index), great care needs to be taken when task demands are not clearly measurable [McKendrick and Cherry, 2018]. Another challenge with questionnaires is that they can only be applied after a task and not during the actual task. Hence, participants only recall the situation which might influence results (recall and post-rationalization bias, [Paxion et al. 2014]). Detailed reviews of workload measures are provided by Butmee et al. [2019] and Paxion et al. [2014].

This case study uses ocular measures to assess workload because they offer continuous monitoring of workload during the experiment. They are based on the fact that, when processing visual information, the eye changes its behavior. For example, the pupil diameter has been shown to increase with added workload [Kramer, 1990]. Recently, the index of pupillary activity (IPA, [Duchowski et al. 2018]) and measures based on micro-saccades Krejtz et al. (2018) are still in early stages. Another eye measure that has been proven to deliver valid results is the index of cognitive activity (ICA, [Marshall 2003]). It is based on the frequency spectrum of the recorded pupil diameter. While the ICA is proprietary, Palinko et al. (2013) define the mean pupil diameter change (MPDC) and mean pupil diameter change rate (MPDCR) as measures for workload. According to their results, especially the latter measure is suited for assessing workload changes during intervals of several seconds. Besides the pupil diameter, blink behavior has been used for assessing workload. Blink latency [Kramer 1990] — the time it takes from the start of an event to the first blink — has been shown to increase with increasing workload. It is based on the theory that with a blink, the human has finished processing the current event. Further, blink rate and blink duration have been shown to be viable measures for workload [Benedetto et al. 2011; Kramer 1990]. However, blink duration seems to be more accurate than blink rate, according to Benedetto et al. [2011]. One aspect that needs to be controlled when using ocular measures is the environmental lighting and its influence on the data. If the lighting is not controlled, it can distort data by introducing additional effects. For example, a dilated pupil due to low lighting.

8.2.2   (Directional) Take-over requests

There is an extensive body of research on in-vehicle warnings in general and take-over requests in detail. In a 2015 crowd-survey with 1692 respondents, Bazilinskyy et al. [2018] found that visual take-over requests on the dashboard are among the most preferred ones. Many researchers used such visual take-over requests during studies - either as main research objective - or as support for other modalities. In many studies, they are used as icons on the center stack [Naujoks et al. 2014; Telpaz et al. 2015] or on the instrument panel [Gold et al. 2013; Lorenz et al. 2014; Louw and Merat 2017; Melcher et al. 2015; Meschtscherjakov et al. 2016; Miller et al. 2014; Mok et al. 2017; Radlmayr et al. 2014]. Regardless of position, these studies used simple icons to indicate a TOR without providing any additional information during situation assessment. These icons often have the design of traditional color-coded warning icons.

To support gaining situation awareness, Rezvani et al. [2016] tried to solve this issue by providing 2.5D illustrations of the surroundings with the intent to communicate the internal and external awareness of the vehicle to the driver during a control transfer. They showed that such visualizations can improve among others, performance after a TOR and situation awareness. Also, displaying icons as a TOR on a head-up display (HUD) has been used to successfully support take-over maneuvers [Kim et al. 2017]. Schwarz and Fastenmeier [2018] went one step further and conducted a study displaying spatially referenced icons on an AR-HUD. Their results and those of others indicate that AR-HUDs perform better than traditional HUDs [Langlois and Soualmi 2016; Weißgerber 2019]. However, presenting information on a HUD display might led to perceptual tunneling [Tönnis et al. 2006] or might even interfere with driver's perception of the surroundings [Kim et al. 2017; Walch et al. 2015].

Wulf et al. [2015] pursue another approach and showed that bringing non-driving related tasks closer to the vehicle's environment (e.g. by providing a video stream next to the non-driving related task) can increase SA but not necessarily driving safety. Related, results of Seppelt and Lee [2007] indicate that providing drivers with a continuous stream of visual information during driving might be more helpful than only punctual information. However, results of Buchhop et al. [2017] argue against that. Their results indicate that providing such continuous stream of information does not necessarily improve driver's reactions in case of a critical event.

A form of visual take-over requests that provides an additional directional information, going beyond a one-dimensional alert, is the use of ambient light. LEDs and LED strips that are placed in the peripheral view of the driver have been shown to support take-over by providing ambient light cues [Borojeni et al. 2016; Löcken et al. 2015]. However, current approaches have not been tested with complex traffic scenarios where an increased amount of spatial information needs to be communicated to the driver. There have also been investigations using directional audio cues [Zarife, 2014], vibro-tactile seats, [Petermeijer et al. 2017a; Telpaz et al. 2015], and shape-changing steering wheels [Borojeni et al. 2017] to indicate a direction or direct drivers' attention. It has been shown that such approaches can shorten response time and can be efficient for warnings. However, they struggled with correctly indicating directions and if the system points towards the danger or towards the way to choose.

As prior research shows, there has been a plethora of investigations in visual and directional take-over requests. Some struggle with the limited capability of the technology to encode information and some with the actual meaning of the directional cue (danger or safe way). Primary objective of our research is to assess if a visualization that shows the surroundings and is presented on a S3D dashboard — thus displayed with binocular depth cues — can support the take-over process. Such icons could convey more information than ambient light and communicate the situation while minimizing room for ambiguity. At the same time, there is no continuous stream of information that still would require monitoring. Also, the spatial nature of the traffic situation could benefit from the presentation in S3D as previous research has shown that the understanding of traffic situations and navigational information is better in S3D compared to 2.5D [Broy, 2015b; Broy et al., 2016; Dettmann and Bullinger, 2019]

8.3   Study Design

Based on the related work and the research objective, the following hypothesis have been formulated:

H1:   A visualization that shows the surroundings as TOR, displayed in 2.5D or S3D, increases mental workload during take-over.

H2:   A visualization that shows the surroundings as TOR increases take-over performance compared to a default warning symbol.

    H2.1:   A visualization displayed at the instrument cluster increases take-over performance.

    H2.2:   A visualization displayed at the focus of attention increases take-over performance.

    H2.3:   A visualization displayed on the gaze path between the road and the focus of attention increases take-over performance.

H3:   A visualization that shows the surroundings as TOR, presented in S3D, leads to better take-over performance than a visualization displayed in 2.5D.

This study had a between-within 4×2 design. Between-subject factor is *type of TOR* and has four levels. The baseline, and three locations where the visualization that shows the surroundings is displayed. Within-subject factor is *type of visualization* and has two levels: 2.5D and S3D.

8.3.1   Procedure

When participants had arrived at the lab, they received and signed the informed consent form. They were then asked to take a seat in the simulator and to adjust the seat to their liking. After that, they had to pass a stereo test and present a valid driver's license from their home country. Failing the test, or having no valid driver's license, led to exclusion. Then, they filled out pre-experiment questionnaires.

Before the experiment started, they put on stereo glasses and the eye tracker was calibrated (c.f. Figure D.1). With the questionnaires finished, a five-minute training drive was performed. Participants also experienced one take-over without any notification to get a feeling of the situation. The end of the training marked the beginning of the experiment. The participants were driven by the automated car along a three-laned road. After around three minutes with 130 km/h, the car detected a lorry in front of it and decelerated to 80 km/h. After a pseudo-random time of 20 – 35 seconds, the lorry performed an emergency braking maneuver, a take-over notification was issued, and the driver had to react. When the TOR was issued, the car was always on the center lane (c.f. Figure 8.1a). Hence, the driver could either evade to the left or to right side of the lorry. In front of the lorry was, hidden from the participant and either on the left or right lane with a distance of 25 m, an occluded vehicle. The participant also had to maneuver around the vehicle. Note that the hidden vehicle is only visible after the TOR had been issued (c.f. Figure 8.1b). Also, a full emergency brake was hardly possible, due to a vehicle that is driving 40 m to 60 m behind the ego car. The tailing vehicle was always visible in a rear-view mirror. A full brake without any evasive maneuver would have led to a rear-end collision. Figure 8.2 illustrates a single TOR scenario. After the first critical situation, participants were instructed to drive along and turn on the automation again. In total, they experienced 8 such take-over situations. Either they experienced first 4 take-overs with a notification displayed in S3D and then 4 in 2.5D or vice versa.

Every time the automation system was enabled, participants were instructed to focus on the non-driving related task. Here, we chose the n-back task [Reimer et al. 2014]. Participants saw a number on the dashboard. This number changed every 1.5 s to another number between 1 and 10. If the current number was similar to the second-to-last number (n = 2), they had to press enter on a keypad. The keypad was located on the armrest to their right. This task was chosen because it requires constant attention and by that, keeps the participant from looking on the road — it simulates or enforces disengagement from the driving task and keeps participants out of the loop.

(a)   (b)

Figure 8.1 Illustration of the driving scenario. Figure 8.1a shows the road just before the TOR. Figure 8.1b a moment after the TOR was issued with the hidden vehicle visible

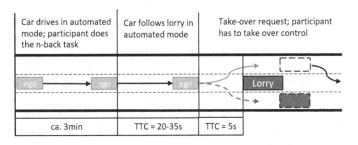

Figure 8.2 Model of a critical situation with timings. Participant (yellow) has to take-over control and evade a hard-breaking lorry and a vehicle hidden behind it

After the first four and second four TORs, participants had to fill out post-experiment questionnaires. That concluded the experiment. Participants had the chance to win 50 Euros. On average, the total experiment lasted 40 minutes.

8.3.2   Driving Simulator

The hardware of the environment simulation and the buck was configured as described in Chapter 4. The setup included head tracking and a rear-view mirror. In addition, there were three Samsung Galaxy S4 smartphones on the sides of the dashboard that acted as markers for the eye tracking system. The setup is illustrated in Figure 8.3.

In this SAE level 3 automation system, the car was able to hold the lane, accelerate, and decelerate. It did not overtake leading vehicles on its own. The driver needs to be fallback ready all the time. An automation indicator communicated the current state of the system (c.f. Figure 8.6a).

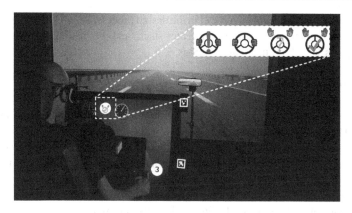

Figure 8.3 Laboratory environment with markers for eye tracking and rear-view mirror

8.3.3 Pre-study: Take-over Notification

Design, Sample, and Procedure
The objective during the design process of the visual take-over notification was to communicate the current traffic situation, to make it easy to understand and perceive, to indicate dangerous objects, and to give a recommendation to the driver on how to handle the situation.

Because no previous literature on the design of such take-over requests was available, we decided to perform a pretest with initial drafts. These drafts have been inspired by in-car visualizations that show traffic situations. Among others, we investigated navigational systems and advanced driving assistant systems (e.g. semi-automated vehicles, lane keeping assist systems, adaptive cruise control).

In this pre-study, we invited N = 5 (2 female, 3 male) participants. They were students and staff from the local technical university (24 - 36 years old; M = 30, SD = 6.52). All had a valid driver's license and drive their car to work (or university) every day.

Participants signed a consent form at the beginning of the study, and we explained the objective of the study to them. We encouraged them to think aloud during the whole process. They took a seat in the driving simulator and enabled automation (SAE Level 3). The car drove on a preliminary version of the scenario that has been used in the actual study (c.f. Section 8.3.1). In regular intervals, participants had to take over control. Each participant was presented each draft of the TOR on each location. The drafts are illustrated in Figure 8.4. For this pre-study, low-fidelity 2.5D graphics were used to avoid overhead in the design process. We presented the drafts on the upper center stack, the lower center stack, and the instrument cluster. All participants saw all drafts on all locations.

Drafts of Visualizations
The first variant, named *abstract*, was a low-poly illustration of the current traffic situation showing important elements in uniformly colored cubes (c.f. Figure 8.4a). The second variant, *photo-realistic*, was the opposite. It details the actual environment and entities photo-realistically (c.f. Figure 8.4b). Note that distances and angles between environment simulation and icon have been preserved. Hence, the hidden vehicle is barely visible. Here, the icon was elliptical because it was based on a screenshot from the environment whereas the others were based on a screenshot from a 3D model. The final variant was a mixed version where the road was textured, and the traffic participants were represented by 3D models of vehicles (c.f. Figure 8.4c). Further, variant *mixed* and *photo-realistic* featured a warning sign highlighting the hidden vehicle. Also, we opted to increase the differentiation between safe and unsafe objects by also coloring the lorry and the hidden vehicle in red. All variants had one version where the system also provided a recommendation in form of an arrow pointing away from the hidden vehicle. In variant *mixed* and *photo-realistic*, the arrow was colored green (based on the color code known from traffic lights) to emphasize the goodness of the recommendation. In the *abstract* version, we did not color the arrow to keep visual complexity low and only highlight the ego car (green) and the hidden vehicle (red). Still we opted for a natural lighting with shadows for aesthetics and to support depth perception. The recommendation was an arrow that pointed towards the safe lane.

       (a) Abstract   (b) Photo-realistic   (c) Mixed

Figure 8.4 Various design samples of the take-over notification with and without recommendations

Results
Every time a take-over was performed, the simulation stopped, and participants were asked to provide their opinion on the visual warning. All participants agreed that the visual warning was a good idea. 3 out of 5 thought that the abstract icon is easy to understand after the first time but not very visually pleasing, especially the lack of

details on the road. One participant was even unsure if the gray flat plane represented the road. On the contrary, participants agreed that the photo-realistic icon could look good. 2 of them mentioned that it could be problematic if more details have to be conveyed. All mentioned that the ego vehicle (here visualized as a vehicle with a green circle below it) should be visualized less ambiguous (P1: "Is it me or is it another driver?"). 4 out of 5 preferred the mixed condition as it presents a balanced view of important information and level of detail. Regarding the navigational arrow, 4 out of 5 said that they would be skeptical of such information. When asked if the icon motivated them to check the rear-view mirror, all responded with "no". One stated that she would "always and ever double check such recommendations — Google Maps makes errors, too.". That agrees with previous research stating that recommendations can impair performance because of humans first confirm the recommendation [Endsley, 2017]. 4 of them mentioned that the icon on the center console could be larger to utilize the available space.

In summary, the version *mixed* was preferred by participants because it provides a good level of detail. Also, before recommendations can be included in such icons, a certain level of trust is necessary. Visualizations should present basic information and only necessary information. The ego-car must be easily identifiable and important areas or details can be highlighted. Dangerous items should be prominently marked.

Figure 8.5 Final take-over notification

Final TOR Visualization
General Design. Based on these findings, we redesigned the TOR. Figure 8.5 shows the result. For the final TOR, we decided to remove the recommendation, avoiding the pitfall of confirmation and trust. A proper investigation would require treating this as an independent variable (with and without recommendation) and further, measure trust.

To better highlight the ego-car, we took inspiration from navigational systems which depict the current location of the ego-vehicle in an arrow-like shape. Hence, the final TOR features such an arrow. Objects or locations that require attention are colored red (hidden vehicle, attention signs, leading vehicle). Following participants' feedback from the pretest, the size of TOR-Large was scaled up with a factor of 1.36. Sizes of objects on the dashboard are listed in the electronic supplementary material D.1. The position of the attention sign was reconsidered. In the end, it was positioned so that it is better visible but still close to the hidden vehicle. Also one sign was added to remind participants that there is a tailing vehicle which is an additional potential source of danger. The visual take-over was accompanied by a siren-like 1.7 s long audio notification that increases in frequency.

Location. When driving with enabled automation, drivers get out-of-the-loop [Endsley and Kiris, 1995]. Time-critical visual info might be missed because the focus of attention is somewhere else (e.g. on the smartphone or on a screen integrated in the center console) and not on the location where it is displayed. A large dashboard can be used to display warnings at the current focus of attention rather than at the traditional place where warnings appear in many cars (the instrument cluster). Going even further, it could eliminate all distracting visual elements and just prominently display the current important warning. Hence, these locations were chosen to study if and how a large display can support take-over maneuvers by displaying warnings. In our prototype and setup, we enforce participants to look at a certain area by means of the non-driving related task. Hence, we can easily estimate the focus of attention. In a real vehicle this could be done by (eye) tracking technology in the future. Figure 8.6 illustrates these locations. The instrument cluster, the focus of attention, large and isolated, or — as a control condition — a traditional warning sign on the instrument cluster.

Stereoscopic 3D. The notifications were displayed respecting the zone of comfort previously defined. The perspective of the icons appears to be different in these illustrations. However, when viewed with an off-axis projection in S3D, all have the same perspective and orientation (road of the icon perpendicular to the road of the environment simulation). Disparity values can be found in the electronic supplementary material D.1.

8.3.4   Measures

During this experiment, we gathered three types of data. First, subjective measures about the experience of participants. Second, ocular measures to get insights about workload. Third, items indicating take-over performance. The following section will

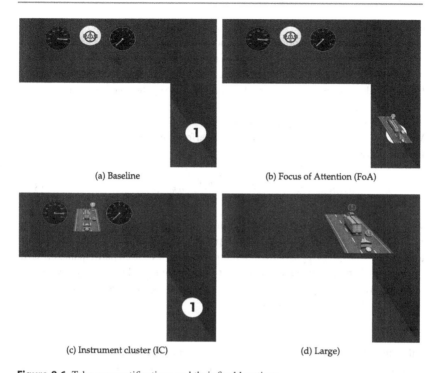

(a) Baseline

(b) Focus of Attention (FoA)

(c) Instrument cluster (IC)

(d) Large)

Figure 8.6 Take-over notifications and their final locations

outline these categories and items. All questionnaires can be found in the electronic supplementary material D.7.

Note that each participant experienced four take-overs in the 2.5D condition and four take-overs in the S3D condition. For each condition and if applicable, we take the arithmetic mean of the four take-overs to control for a habituation effect and to stabilize values.

Subjective Measures
To assess the user experience of people, we asked participants to fill out the User Experience Questionnaire (UEQ, [Laugwitz et al. 2008]). Because driving in a driving simulator can induce simulator sickness (as has previously been shown), participants were required to fill out the simulator sickness questionnaire (SSQ, [Kennedy et al. 1993]). Both questionnaires are used after each condition. That means, each participant filled out the UEQ and SSQ twice — after the first four take

overs (in either 2.5D or S3D) and after the second four take overs (after 2.5D or S3D, respectively). Both questionnaires were administered via a tablet. Participants did not leave the driving simulator but took of glasses. We also measured gaze behavior with respect to five area of interests. These areas of interest covered the front screen (the road), the rear-view mirror, and the areas taken up by the take-over notifications.

Non-driving related task performance
We also measured non-driving related task performance by calculating the *nBack-Rate* as follows:

$$nBackRate \; = \; \frac{n_{Ratio} - N_{min}}{N_{max} - N_{min}}, \;\; n_{Ratio} \; = \; \frac{n_{correct}}{n_{Total}} \qquad (8.1)$$

with N_{min} and N_{max} being the total minimum and maximum ration of the sample, $n_{correct}$ the number of correct responses per participant, and n_{Total} the number of total required responses per participant. It represents how often participants pressed the button when the second to last number was the same as the one they saw at that moment. We use this a control variable. A high value suggests that participants were engaged in the task and vice versa.

Workload
All ocular data was gathered using an Ergoneers DIKABILIS Professional eye tracker. The eye tracker was mounted on the S3D glasses (c.f. electronic supplementary material D.2). Camera adjustment and final calibration was done for each participant by the experimenter following the procedure proposed by the manufacturer. We started measurement when automation was enabled first and ended with the end of the track. For measures during take-over, we recorded data 7 seconds after the TOR was issued as post-hoc analysis showed that all participants had then finished the take-over maneuver.

MPDC - Mean Pupil Diameter Change   To assess workload, Palinko et al. (2013) propose the mean pupil diameter change *MPDC*. It is calculated as

$$MPDC = \frac{\sum_{i=0}^{n_{TOR}} (MPD_i - MPD_{total})}{n_{TOR}} \qquad (8.2)$$

whereas n_{TOR} is the total number of take overs, MPD_i is the mean pupil diameter (as measured by, for example, an eye tracking device) of the i-th take-over, MPD_{total} is the average mean pupil diameter, and n_{TOR} is the total number of observed take overs. In other words, it tells us if the pupil diameter of a participant was different to the average of all participants. A lower $MPDC$ indicates a lower workload [Palinko et al. 2013]. Note that with this measure, the overall workload cannot be determined but it enables comparisons between different groups.

MPDCR - Mean Pupil Diameter Change Rate The second measure considering the pupil diameter is the mean pupil diameter change rate $MPDCR$ [Palinko et al. 2013]. It is calculated as

$$MPDCR = \frac{\sum_{i=0}^{n_{TOR}} \frac{\sum_{j=1}^{n_{TOR}-1} F'(t)}{n_{TOR}-2}}{n_{TOR}} \tag{8.3}$$

whereas t represents the time in ms, n_T the total number of measured pupil diameters, n_{TOR} the total amount of take-overs, and $F'(t) = \frac{F(t+h)-F(t-h)}{2h}$ the first derivative indicating change rate. $F(t)$ represents the mean pupil diameter MPD at time t, and h the time between measurement points. A positive MPDCR represents pupil dilation during an event and by that increased workload and a negative pupil contraction (and decreased workload). The authors mention that this measure is especially suited for detecting changes of workload in short intervals smaller than several seconds.

MBDC - Mean Blink Duration Change In accordance with the calculation of the mean pupil diameter change, we calculate the mean blink duration change MBDC as follows:

$$MBDC = \frac{\sum_{i=0}^{n_{TOR}} (MBD_i - MBD_{total})}{n_{TOR}} \tag{8.4}$$

whereas MBD_i is the i-th blink duration, MBD_{total} the average blink duration of all participants and TORs, and n_{TOR} the number of take-overs per participant. MBDC represents the difference of one participant's mean blink duration regarding the total average blink duration. A positive MBDC means that the blink duration during an event is larger than the duration during the rest of the drive and vice versa. A smaller MBDC indicates a higher workload, probably due to the visual system wanting to minimize time where it does not see anything (Kramer 1990; Marquart et al. 2015). Blink duration can be obtained using, for example, an eye-tracking device.

Blink Latency Blink latency is defined as the time between the end of a take-over and the time the first blink was detected.

$$BL = \frac{\sum_{i=0}^{n_{TOR}} t_{TOR_i} - t_{BlinkDetected_i}}{n_{TOR}} \qquad (8.5)$$

whereas n_{TOR} is the number of TORs, t_{TOR_i} is the time the i-th TOR was issued, and $t_{BlinkDetected_i}$ is the time of the first blink after the i-th TOR. A larger blink latency, meaning a larger time between the issue of a TOR and the first blink, indicates increased visual workload. Research hypothesizes that the visual system wants to get as much information as possible in the critical situation and hence, inhibits blinks Benedetto et al. (2011), Kramer (1990), Recarte et al. (2008). If a participant did not blink for 7 seconds after the take over notification was issued, their $t_{BlinkDetected}$ was considered to be 7 seconds. Again, this measure allows us to compare groups but not to make assumptions about high or low workload in general.

Take-over performance

The third category of measures is about take-over performance. We measured the *number of safe take-overs* per participant. A take-over was considered safe when two conditions apply: first, there was no collision with any foreign object (railing or vehicles). Second, the participant maneuvered the vehicle towards the less risky side of the road.

We further measured *motor reaction time*. It is defined as the time it took participants to turn the steering wheel more than 5° towards either side or hit a pedal more than 10%. The latter measure was adopted from Gold et al. [2013]. Gold et al. also propose a 2° rule for the wheel turn angle. Due to backlash (or play) and inaccuracies of our steering wheel, we had to increase this value to 5° per direction after our pretests. Turning the steering wheel for 5° to either side changes the forward vector of the virtual vehicle by 2° (the wheel has 3° play). For each participant, we calculated the arithmetic average of the four motor reaction times. We calibrated the steering wheel for each participant using the control panel of the manufacturer[1]. Steering 10% has an influence of v = 1.420 km/h per 10 seconds (SD = 0.014 km/h; averaged over N = 5 trials) on vehicle velocity.

The third measure for take-over performance is the time it took participants to first look at the road [Zeeb et al. 2016]. It is defined as the duration $\Delta t = t_{FirstLookAtRoad} - t_{TOR}$ with $t_{FirstLookAtRoad}$ being the time it took participants to look at the road and t_{TOR} describing the time the TOR was issued. Again, we took the arithmetic average of the four Δt's.

---

[1] https://support.thrustmaster.com/en/product/txracingwheelleather-en/, Thrustmaster - Technical support, 2020-09-24

8.3.5   Sample

52 participants (34 male, 18 female; 19 - 63 years, M = 31.9, SD = 10.6) took part in the experiment. They were recruited using convenience sampling (Facebook groups, personal contact, mailing lists). Hence, most of them were students and staff of the local university. All had a valid driving license, normal or corrected vision, and passed a stereo vision test [Schwartz and Krantz, 2018]. 36 stated that they had previous experience with stereoscopic 3D displays. 30 participants reported that they had no experience with driving simulators. 10 of them have been in a driving simulator once. 7 have been in one between 1 and 5 times. 5 participants have been in a driving simulator more than 5 times. The final sample has a mean MSSQ score of M = 8.45 (SD = 7,06). Participants of the prestudy were excluded from participation.

8.4   Results

All numerical data was analyzed with R 3.6.1 (afex 0.24.1, car 3.0.3, bestNormalize 1.4.0, emmeans 1.4, doBy 4.6.2, and fBasics 3042.89). An α-value of 0.05 was used as significance criterion when necessary. Outliers removed above or below 2 times the inter-quartile range [Elliott and Woodward, 2007].

Eye tracking data was analyzed using D-Lab 3.51[2]. We ran built-in pupil recognition algorithms. A manual screening of all take-over process from 5 s before the take-over to 7 s after the take-over was performed to fill potential gaps in detection or correct false classifications but also to confirm that all participants were focusing on the n-back task prior to the take-over request.

Data was analyzed using mixed ANOVAs. Normality of the residuals was assessed using Shapiro-Wilk tests and QQ-plots [Yap and Sim, 2011]. Homoscedasticity was tested using Levene's test [Brown and Forsythe, 1974]. If not mentioned otherwise, no corrections were necessary as Mauchly's test [Mauchly, 1940] for sphericity was not significant with p > 0.05.

Average experiment duration was 40 minutes. There were no dropouts. Every participant had to perform 4×2 take-overs (4 TORs per condition, 2.5D and S3D). By that, a total of $52 \times 8 = 416$ take-overs are available for analysis.

---

[2] http://ergoneers.com, 2020-09-24

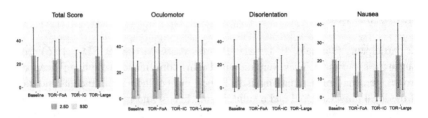

Figure 8.7 Mean values and standard deviations of the simulator sickness symptoms (lower is better; O = Oculomotor [0;159.18], N = Nausea [0;295.74], D = Disorientation [0;292.32], TS = Total Score [0;235.62])

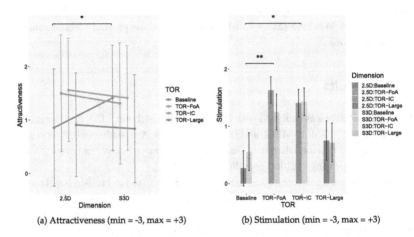

(a) Attractiveness (min = -3, max = +3)   (b) Stimulation (min = -3, max = +3)

Figure 8.8 Results of the UEQ *Attractiveness* and *Stimulation*. Both showed significant differences (higher values are better; significance codes: ***: p < 0.001, **: p < 0.01, *: p < 0.05)

8.4.1   Simulator Sickness Questionnaire

Figure 8.7 lists data of the SSQ for the subscales and total score. Data of the SSQ was approximately normal distributed. Mixed ANOVAs did not reveal any main or interaction effect ($F(3, 48) <= 1.59$, $p > .197$, $\eta_p^2 < 0.047$). The complete set of statistical results can be found in the electronic supplementary material D.3. By that, we can assume that simulator sickness experience was the same among all groups.

8.4.2   User Experience Questionnaire

No subscale of the UEQ followed an approximation of the normal distribution. Hence, we applied an ordered quantile normalizing transformation.

Results of a mixed ANOVA on the data of the transformed subscale *Attractiveness* indicate an interaction effect of TOR and dimension ($F(3, 48) = 4.30$, $p = .009$, $\eta_p^2 = .21$). Figure 8.8a illustrates this effect. Post-hoc analysis using Tukey-corrected pairwise comparisons indicate that there is a significant difference between the S3D-baseline ($M = 0.86$, $SD = 1.10$) and 2.5D-baseline ($M = 1.41$, $SD = 0.99$) favoring the S3D condition ($t(48) = 3.270$, $p = .0386$, Cohen's $d = 0.52$). That tells us that participants of the baseline condition liked the S3D effect more than presentation in 2.5D mode.

On the scale *Stimulation*, there is a main effect of TOR (c.f. Figure 8.8b, $F(3, 48) = 5.75$, $p = .002$, $\eta_p^2 = .26$). Post-hoc analysis indicates a significant difference between TOR-IC and baseline ($t(48) = 3.193$, $p = .0129$, Cohen's $d = 0.997$) as well as TOR-FoA and baseline ($t(48) = 3.299$, $p = 0.0096$, Cohen's $d = 0.010$). That tells us that TOR-IC ($M = 1.416$, $SD = 0.850$) and TOR-FoA ($M = 1.442$, $SD = 0.983$) are perceived more stimulating (or exciting) than the baseline condition ($M = 0.413$, $SD = 1.138$).

There were no other main or interaction effects. Neither for these, nor for the other scales

Figure 8.9 contextualizes the results of the UEQ using the classification system introduced by Schrepp [2017]. It contains a five-point rating system going from bad to excellent. Note that it only considers the mean values. Figure 8.9a illustrates

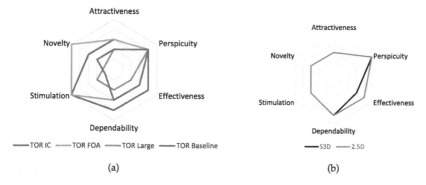

(a)   (b)

Figure 8.9 Results of the UEQ in context of the rating scheme by Schrepp et al. [2017] (from inside to outside: bad - below average - above average - good - excellent)

the classification according to take-over notification. Figure 8.9b groups results by dimension. The electronic supplementary material D.4 contains descriptive statistics and results of ANOVAs of the UEQ.

8.4.3   Glance Behavior

Number of glances
Data representing the number of glances was approximately normal distributed. Results of a mixed ANOVA suggest interaction effect of TOR and dimension ($F(3,48) = 3.30, p = .028, \eta_p^2 = .17$). Post-hoc analysis using Tukey-corrected pairwise comparisons indicate that TOR-IC in S3D received more glances than all other combinations except TOR-Large in S3D ($t(48) > 3.421, p < 0.05$, Cohen's $d > 0.77$). Figure 8.10 illustrates the data.

In addition to that, results indicate a main effect of TOR ($F(3,48) = 4.88, p = .005, \eta_p^2 = .23$) and also of dimension ($F(1,48) = 5.88, p = .019, \eta_p^2 = .110$). We report them

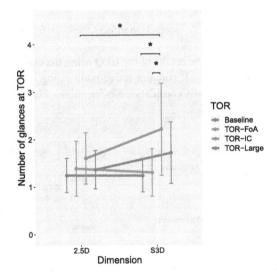

Figure 8.10 Results on the number of glances at the single TORs. S3D-TOR-IC received significantly more glances than all other TORs except S3D-TOR-Large (significance codes: ***: p < 0.001, **: p < 0.01, *: p < 0.05)

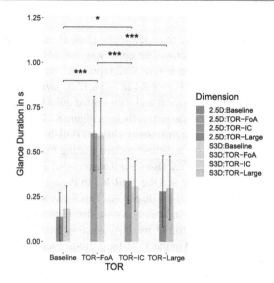

Figure 8.11 Results on the glance duration at the single TORs. TOR-IC and TOR-FoA accumulated more glance time than baseline (significance codes: ***: p < 0.001, **: p < 0.01, *: p < 0.05)

for the sake of completeness but they are represented by the mentioned interaction effect. There were no other main or interaction effects. Complete test results and descriptive statistics can be found in the electronic supplementary material D.5.

Glance duration

Figure 8.11 illustrates the glance duration accumulated during the take-over process by the single take-over notifications per condition. Data representing glance duration was approximately normal distributed. Results of a mixed ANOVA suggest a significant main effect of TOR ($F(3, 48) = 19.30, p < .001, \eta_p^2 = .55$). Post-hoc tests using Tukey-corrected pairwise comparisons indicates that the baseline TOR was looked at less than TOR-FoA $t(48) = -7.394, p < .0001$, Cohen's $d = 2.55$). Also, TOR-IC received in total shorter glances than TOR-FoA ($t(48) = -4.603, p = 0.0002$, Cohen's $d = 1.61$). Similarly, TOR-Large received shorter glances than TOR-FoA ($t(48) = 5.194, p < .0001$, Cohen's $d = 1.58$). In summary, participants looked longer at TOR-FoA than at the other TORs. In addition to that, the baseline accumulated a shorter glance duration than TOR-IC ($t(48) = -2.791, p = 0.037$, Cohen's $d = 1.26$). There was no main effect of dimension ($F(1, 48) = 0.08, p = .783, \eta_p^2 = .01$) or an interaction effect ($F(3, 48) = 0.61, p = .612, \eta_p^2 = .04$).

Gaze patterns

This section provides a qualitative overview of the glance behavior per participant. Figure 8.12 shows the gaze patterns. The upper row for 2.5D, the lower row for S3D. One line in one sub-plot in Figure 8.12 represents the gaze pattern of one participant. For the baseline (c.f. Figure 8.12a and 8.12e), gaze mostly remained at the location of the n-back task (green) and then shifted quickly to the road (grey). Few participants took a glance at the mirror (beige). Figure 8.12b and 8.12f show the gaze behavior for the TOR at the instrument cluster (TOR-IC, blue). Again, there are few glances at the mirror but a distinct pattern starting from the n-back task to the IC and then to the road. There are also several glances that indicate an alternating gaze, switching between the road and the TOR-IC. In Figure 8.12c and 8.12g, the gaze behavior for TOR-FoA (green) is displayed. This figure shows longer glances at the location of the TOR-FoA, indicating a gaze that remains there for a period of time and then quickly shifts to the road. Again, few participants checked the mirror or the instrument cluster. For the TOR-Large, illustrated in Figure 8.12d and 8.12h, rather short and non-repeating glances are visible. Participants gaze neither remained for a long time at the TOR-Large (orange), nor at the n-back task. They also rarely looked at the mirror or the instrument cluster.

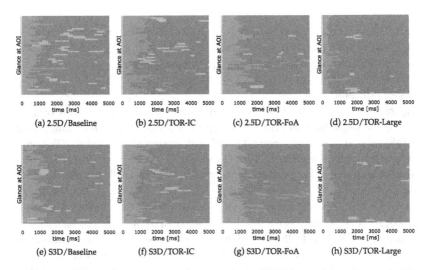

Figure 8.12 Scarf plots of participants gaze behavior. First row: 2.5D. Second row: S3D (Baseline/TOR-IC = blue, TOR-FoA = green, TOR-Large = orange, rear-view mirror = beige, front = grey)

8.4.4 Non-driving related task Performance: n-Back-Rate

Figure 8.14 shows the n-back rate per participant and condition. Data did not follow an approximation of the normal distribution. An ordered quantile normalizing transformation (Peterson and Cavanaugh 2019) led to an approximate normal distribution. Using mixed ANOVAs, we could not show any main effect (TOR: $F(3, 48) = 0.92$, $p = .439$, $\eta_p^2 = .05$, dimension: $F(1, 48) = 1.37$, $p = .248$, $\eta_p^2 = .03$) or interaction effect ($F(3, 48) = 0.43$, $p = .730$, $\eta_p^2 = .03$) on the transformed data. That tells us that no group performed better or worse than any other group in the n-back task.

8.4.5 Workload

Blink latency
Figure 8.13 describes the blink latency per TOR and type of visualization. 15 participants did not blink. For them, blink latency was set to the duration of the observed interval with BL = 7 s. Data was not approximately normal distributed. Hence, an ordered quantile normalizing transformation was applied.

A mixed ANOVA on the transformed data provided no evidence for a main or interaction effect (TOR: $F(3, 48) = 1.66$, $p = .187$, $\eta_p^2 = .09$; dimension: $F(1, 48) = 0.18$, $p = .669$, $\eta_p^2 < .01$; interaction: $F(3, 48) = 0.52$, $p = .668$, $\eta_p^2 = .03$) on the transformed data. That tells us that the blink latency did not differ between groups and indicates that workload did not differ.

MBDC: Mean blink duration change
The statistics on MBDC was calculated without the 15 participants that did not blink in the observed interval. There was no logical or natural substitute value. Hence, analysis was performed without those 15 items. Data representing the mean blink duration change was approximately normal distributed. A mixed ANOVA on the untransformed data did suggest a main effect of dimension ($F(1, 37) = 4.37$, $p = .043$, $\eta_p^2 = .11$). It suggests that the MBDC in the 2.5D condition ($M_{2.5D} = -107.09$ ms, SD = 52.06) is significantly higher than in the S3D condition ($M_{S3D} = -114.82$ ms, SD = 57.59). That tells us that the blink duration during events in the 2.5D condition was higher than in the S3D condition. That, in return, indicates that workload in the S3D condition was higher than in the 2.5D condition. There were no other main ($F(3, 37) = 1.45$, $p = .245$, $\eta_p^2 = .10$) or interaction effects ($F(3, 37) = 0.28$, $p = .841$, $\eta_p^2 = .02$). Figure 8.15 provides an overview of the data.

Figure 8.13 Blink Latency
in ms (higher values
indicate more workload)

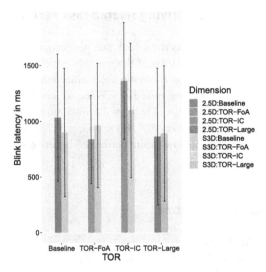

Figure 8.14 Descriptive
statistics of the
untransformed nBackRate
in % (higher values indicate
better performance)

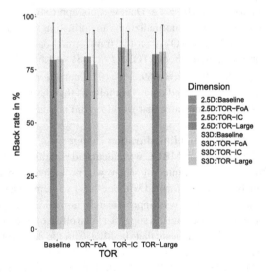

MPDC: Mean pupil diameter change

Data representing the MPDC did not follow an approximation of the normal distribution. An ordered quantile normalizing transformation approximately restored normal distribution. A mixed ANOVA on the transformed data did not suggests any main effect of TOR ($F(3, 48) = 1.13$, $p = .347$, $\eta_p^2 = .07$) or dimension ($F(1, 48) = 0.00$, $p = .992$, $\eta_p^2 < .01$). There was also no interaction effect ($F(3, 48) = 1.22$, $p = .314$, $\eta_p^2 = .07$). Figure 8.16 lists descriptive statistics of the MPDC.

MPDCR: Mean pupil diameter change rate

Similarly to MPDC, the MPDCR data was not approximately normal distributed. Figure 8.17 lists descriptive statistics of the data. Again, transforming data using an ordered quantile normalizing transformation led to approximate normal distribution. Results of a mixed ANOVA did not indicate any main or interaction effect. Neither TOR ($F(3, 48) = 2.58$, $p = .064$, $\eta_p^2 = .14$), nor dimension ($F(1, 48) = 1.30$, $p = .259$, $\eta_p^2 = .03$) indicated significant differences. There was also no significant interaction effect ($F(3, 48) = 0.22$, $p = .882$, $\eta_p^2 = .01$). That tells us that the amount a pupil contracted or dilated per participant during events did not differ between groups. In return, that indicates that workload did not differ.

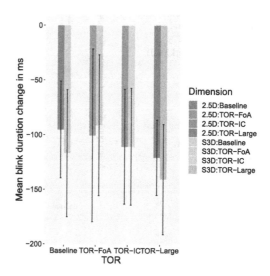

Figure 8.15 Mean Blink Duration Change (MBDC) in ms (> 0: blink duration during an event is larger than the duration during the rest of the drive and vice versa; smaller MBDC indicates higher workload)

Figure 8.16 Mean Pupil
Diameter Change (MPDC)
in pixel (> 0: pupil diameter
during an event is larger
than the diameter during the
rest of the drive and vice
versa; higher MPDC
indicates higher workload)

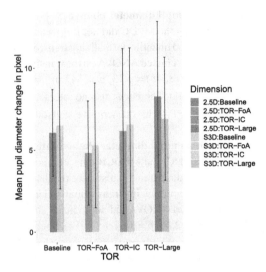

Figure 8.17 Mean Pupil
Diameter Change Rate
(MPDCR) in pixel (> 0:
pupil dilated per event more
than it contracted and vice
versa; higher MPDCR
indicates higher workload)

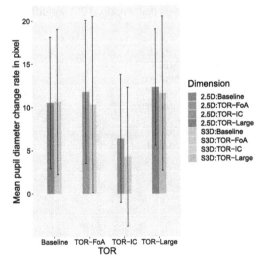

Figure 8.18 Motor reaction time in [ms] (lower is better)

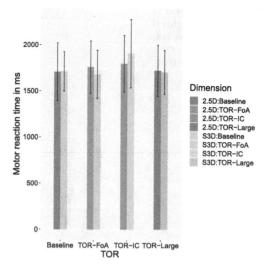

Figure 8.19 Time to first glance at the road in ms (lower is better; significance codes: ***: p < 0.001, **: p < 0.01, *: p < 0.05)

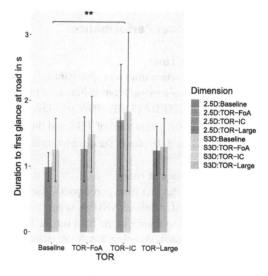

Figure 8.20 Number of safe take-overs (higher is better, min = 0, max = 4, significance codes: ***: p < 0.001, **: p < 0.01, *: p < 0.05)

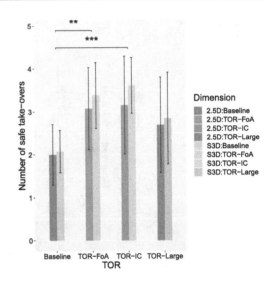

8.4.6   Take-over Performance

Motor Reaction Time

Data on motor reaction time was approximately normal distributed. Neither dimension nor TOR led to significant differences in participants' motor reaction time (mixed ANOVA, TOR: $F(3, 48) = 0.89, p = .452, \eta_p^2 = .05$, dimension: $F(1, 48) = 0.03$, $p = .875, \eta_p^2 < .01$, interaction of TOR and dimension: $F(3, 48) = 1.77, p = .165$, $\eta_p^2 = .10$)). Figure 8.18 shows the descriptive statistics.

Time to first glance at road

Data describing the first time participants glanced at the road was approximately normal distributed. A mixed ANOVA suggests a main effect of TOR on the time it took participants to glance at the road for the first time ($F(3,48) = 4.06, p = .012$, $\eta_p^2 = .20$). Post-hoc tests using Tukey-corrected pairwise comparisons pointed out that there are significant differences between the baseline and TOR-IC ($t(48) = -3.341, p = .0085$, Cohen's $d = 0.84$). That tells us that, when the baseline TOR was shown, participants took less time to look at the road compared to TOR-IC — Figure 8.19 illustrates this difference. Results did not uncover a main effect of dimension ($F(1, 48) = 2.18, p = .146, \eta_p^2 = .04$) or an interaction effect ($F(3, 48) = 0.17, p = .918$,

$\eta_p^2 = .01$). Descriptive statistics are listed in the electronic supplementary material D.6.

Number of Safe Take-overs

Figure 8.20 shows results of the take-over performance in form of the number of safe take-over maneuvers. Out of all 416 take-overs, 119 were classified as wrong. 14 were due to the participants making a full stop. By that, they risk a collision or were involved in a rear-end collision. Data was approximately normal distributed. A mixed ANOVA did not uncover a significant interaction effect ($F(3, 48) = 2.22$, $p = .098$, $\eta_p^2 = .12$). Results indicate a main effect of the presented take-over notification ($F(3, 48) = 7.75$, $p < .001$, $\eta_p^2 = .33$). Post-hoc tests (Tukey-corrected pairwise comparisons) indicated, that the baseline and TOR-IC ($t(48)=-4.384$, $p=.0004$, Cohen's $d= 1.706$) as well as the baseline and TOR-FoA ($t(48)= 3.468$, $p=.0059$, Cohen's $d=1.605$) differ significantly. That tells us that a visualization in form of TOR-IC and TOR-FoA led to more safe take-over maneuvers than the baseline condition. The TOR-Large visualization was neither better nor worse than baseline, TOR-IC or TOR-FoA. There was also a main effect of dimension ($F(1, 48) = 4.51$, $p = .039$, $\eta_p^2 = .09$). Here, data indicates that S3D led to more safe take-overs than 2.5D ($M_{S3D} = 2.98$, $SD_{S3D} = 0.960$; $M_{2.5D} = 2.73$, $SD_{2.5D} = 1.06$). That tells us that an S3D visualization led to more safe take-over maneuvers than 2.5D. Note, TOR was a between-subject variable. Each participant experienced only one type of TOR.

8.5 Discussion

In this study, we investigated if visualizations that show the surroundings presented in S3D can support the take-over process. We hypothesized that such a visualization is better than traditional icons that do not convey information about the environment (RQ1). Further, we assumed that the location where such a visualization is displayed, influences take-over performance as well (RQ2). Also, we assumed that communicating the spatial information on a S3D display using binocular depth cues will improve performance (RQ3). Finally, we expected that mental workload is increased by the added visual information (RQ4).

In the following, we are going to discuss the results of the previous chapters. First, we will dive into control variables. Namely, glance behavior, simulator sickness, non-driving related task performance, and user experience. This data gives us an overview what impact the various visualizations had on participants. After that, we will discuss the workload measures followed by the measures indicating take-over performance.

8.5.1   Perception of the system

We did not measure any significant difference in the SSQ data. However, the overall experience of simulator sickness symptoms reported by participants is rather high. While data are similar to results in our previously reported studies, we still have to consider their impact in the overall results. The elevated scores might have masked effects of TOR or type of visualization.

Regarding user experience, few differences were found by the statistical analysis. Participants preferred the S3D version of the baseline condition compared to the 2.5D version. For the other take-over notifications, no difference could be found. We assume that the plasticity of the elements paired with the clean and de-cluttered layout led to this difference. Participants also found TOR-IC and TOR-FoA more stimulating than the baseline TOR. In the definition of the UEQ scales, stimulation is defined as being exciting or motivating. One explanation for this difference is that the participants actively engaged with these notifications. They either switched their gaze to TOR-IC and analyzed it or their gaze remained there for a longer time (in the case of the TOR-FoA). The interface in the baseline condition did not provide any exciting or motivating elements which explains the low values on this scale. One explanation for the absence of a difference between TOR-Large and the baseline is that participants gazed only briefly at TOR-Large and did not dedicate many glances to it. It is possible that they did not fully process the TOR or that they did not pay too much attention to the icon and hence, did not attribute much stimulating or motivating properties to it. There were some noticeable differences in the evaluation of the gaze behavior. A qualitative analysis showed salient differences or patterns between the three take-over notifications. The patterns can be named *switch*, *observe*, and *fly-by*. The *switch*-pattern, demonstrated in TOR-IC, contains many small back-and-forth glances between the road and the instrument cluster. It is likely that these glances were necessary to fully capture all the information of the icon but also maybe to confirm the provided data. In the S3D case, there is also the possibility that a second or third look on the icon was necessary to establish binocular fusion (the fusion process can take up to 200 ms, [Howard and Rogers, 2008]). A high number of glances is characteristic for this pattern (Section 8.4.3). The *observe*-pattern is characterized by the long glance in the beginning followed by the movement of the gaze towards the road. In this case, the TOR was issued, and participants' gaze remained at the location until they had fully processed the icon.

While some took another glance at the TOR after they had looked at the road, the majority did not. The rather long glance duration (Section 8.4.3) and few number of glances Section 8.4.3 are characteristic for this pattern. Finally, the *fly-by*-pattern is demonstrated by the behavior exhibited in the TOR-Large condition. Participants glance quickly shifted from the n-back task to the TOR and then to the road. Glance duration and glance behavior did not show any salient characteristics except a clear movement pattern from n-back task to the road with a low dwell time on the take-over notification (not different to baseline).

Important for the evaluation of the effectiveness of our TOR is how involved participants were in the non-driving related task, which was designed to keep them out-of-the-loop and simulate a non-driving related task. Overall performance of the n-back task was high. In M = 82% (SD = 13%), participants gave correct answers. The absence of any difference between groups indicates that all participants were engaged in the non-driving related task. The n-back task with n = 2 is cognitively demanding which explains the error rate of around 20%.

In summary, the participants of the different conditions had rather similar responses towards the system, considering UEQ, SSQ and the n-back task. There were noticeable differences in the gaze behavior which need to be considered in future applications.

8.5.2 Workload (H1)

RQ1 of this study asked if the workload was impaired by the added visual information compared to baseline. Workload was measured via blink latency, mean blink duration change, mean pupil diameter change, and mean pupil diameter change rate. Using ocular measures for workload provides us with means to assess the workload during the actual take-over processes, avoiding recall and post-rationalization bias.

In blink latency (a higher blink latency indicates higher workload), no differences could be found between the take-over notifications with and without an icon. Similarly, the MPDC (mean pupil diameter change; a higher mean pupil diameter change indicates higher workload), did not differ. Analysis of the MPDCR (mean pupil diameter change rate; indicates amount of dilation/contraction during events; higher indicates more workload) did also not reveal an effect of dimension or type of take-over notification. Only the MBDC (mean blink duration change; higher MBDC indicates increased workload) suggests increased workload in the S3D condition. Here, mean blink duration change during events was significantly lower in the 2.5D condition compared to the S3D condition. While the blink duration during the events decreased for both types of visualization (2.5D and S3D), it decreased

even more in the S3D condition. Note that the overall difference between MBDC in
2.5D and S3D is $M_{2.5D} - M_{S3D} = -107.1$ ms $- (-114.8$ ms$) = 7.7$ ms and with that,
rather small. However, a similar difference of 10 ms was found by Stern and Skelly
[1984] for increased workload . Hence, this can be taken as an indication for a higher
workload in the S3D condition. A closer look at the descriptive statistics shows that
especially TOR-Large in S3D shows a very low average MBDC (-140.87 ms in
S3D and -120.99 in 2.5D). Concluding, S3D seems to have at least some effect on
mental workload represented by mean blink duration change — blinks got shorter.
Possibly due to the fact that blinks influence binocular fusion. Also, it does not seem
to be the case that the added visual information influences workload as we could
not detect significant differences between baseline and the visual TORs.

8.5.3   Type of Visualization and Location (H2)

H2 stated that a visualization that shows the surroundings increases take-over performance. Hence, TOR-IC, TOR-FoA, and TOR-Large should all show better performance than the baseline condition. In our case, that means that the motor reaction
time and the time to the first glance on the road were not worse than baseline. Also,
it suggests that the number of safe take-overs should be higher when using TORs
showing the surroundings compared to baseline.

Motor reaction time did not differ between conditions. That is in line with previous research, stating that motor readiness is quickly available [Zeeb et al. 2016].
Compared to previous research, the time it took participants to make their first look
on the road is slightly higher. Zeeb et al. [2016] mention times between 0.77 s to
0.99 s whereas the mean values in our experiment range from 0.99 s in the baseline-
2.5D condition to 1.84 s in the S3D-TOR-IC condition. In the study of Vogelpohl
et al. [2018], eyes on the road time is between 0.12 s for a no-task condition up
to 1.90 s for a gaming condition. At best, this time is minimal — meaning that
participants react instantaneously and scan the road to gain situation awareness. By
presenting additional visual information to the driver, it was likely that this value
is significantly higher when using a visualization that shows the surroundings compared to the baseline. Interestingly, only the TOR displayed on the instrument cluster
(TOR-IC) led to a significantly delayed first glance to the road. We believe that the
rather small size of the icon paired with the movement of the head and the S3D
visualization led to participants needing a longer time to fully focus on, perceive,
and understand the information before they look towards the road and react. This
is not true for the other conditions with visualizations that show the surroundings.
Here, there was no significant difference regarding the first glance towards the road

compared to baseline. For TOR-Large, this could mean that it is either rather easily to perceive and understand or that participants completely ignored it. However, the non-significant results argue for the latter. For TOR-FoA, one explanation is that the absence of an additional head movement and accompanying refocusing support understanding of the icon and by that take-over performance.

Considering the number of safe take-overs, TOR-Large did not led to significantly different results when comparing it with the baseline take-over notification. While we could argue that the location was unsuitable, it is also possible that our motivation to fully utilize the dashboard and available design space had a detrimental effect. We intended to increase the size of the TOR to improve visibility and understanding of the icon. Results of the pretest motivated size and location. Retrospectively, it could also be the case that the increased size had the opposite effect. Its size could have negatively influenced binocular fusion and understanding in the short (and time-critical) interval. Form the presented data, it is noticeable that TOR-IC and TOR-FoA attracted participants' gaze and led to more safe take-overs whereas TOR-Large received only few and short glances. That suggests that participants tried to perceive it but did not fully process information and hence, most likely reacted based on experience and intuition. Here, performance and gaze behavior indicate that participants actively used the icons and acted according to the information. An interesting observation is that, while it took participants longer to look at the road for the first time in the TOR-IC condition, the number of safe take-overs was still significantly higher compared to baseline. That can be taken as an indication that the added information of the visualization supported the driver in making the correct decision.

Overall, applying a visualization that shows the surroundings did not per se improve take-over performance. We tested them in three cases, but participants did not perform better in all of them but only in two. Hence, we can answer RQ2 in the following way: there were some significant impairments of performance (time to first glance at the road) but also significant improvements (number of safe take-overs) when using a visualization that shows the surroundings. Interestingly, TOR-IC, the condition with the latest first glance to the road, also led to the highest number of safe take-overs. TOR-FoA performed equally good. Location (and potentially size) play a role that needs to be investigated further. The improvements in the number of safe-take overs can be considered a strong motivation for future research and as an indication that icons conveying situational information might foster understanding of traffic situations and by that, increase safety.

8.5.4   Binocular Depth Cues (H3)

RQ3 asked if visual presentation of the current traffic situation using a visualization that shows the surroundings in S3D increases take-over performance. In other words, we measured how often participants chose the safe route instead of the unsafe rout with the hidden vehicle which presented additional danger. Results indicate that binocular depth cues can increase perception of such visualizations: overall, the S3D conditions led to significantly more safe take-overs than the 2.5D conditions. Participants in the 2.5D group handled on average 2.73 situations in a safe manner (max = 4). The participants in the S3D group handled 2.98 maneuvers safely (max = 4). This difference, while small, is significant. By that we can assume that the spatial presentation of navigational information has a positive impact on perception of the icon and upcoming execution of the take-over process. In addition to that, motor reaction time was not influenced by the type of visualization. However, compared to the other conditions, presentation in S3D sometimes led to more glances at the notification and away from the road (e.g. S3D-IC, c.f. Section 8.4.3). Nevertheless, this did not impair take-over performance. On the contrary, performance was still significantly better than baseline (c.f. Section 8.4.6). Despite the fact that the absolute values of the number of safe take-overs is higher in all the S3D conditions, the difference regarding their 2.5D counterpart was not statistically significant. Here, we have to take into account that the effect was overall small with $\eta_p^2 = .09$. That means that, overall, S3D led to more safe-take overs but for each type of TOR, it did not perform good enough to suggest a statistically significant improvement compared to the 2.5D counterpart. By that, we can conclude that there is evidence suggesting that presentation in S3D has the potential to increase perception of visualizations that show the surroundings presented during take-over.

8.5.5   Limitations & Future Work

Results of our study are encouraging. However, generalization is limited.

One limitation is the traffic scenario: a suddenly appearing object and pseudo-randomly stopping vehicles. Such scenarios do not necessarily occur in the wild but are often used in research studies. No participant commented on a lack of realism However, tests with various other scenarios are necessary to confirm these findings. In addition to that, the simulator lacks any movement. Prior research has shown that vehicle movement, especially in curved segments, impacts the decision of drivers during evasive take-over maneuvers [Borojeni et al. 2018]. Also, our warning resembled a stop sign rather than a triangular warning sign which might have participants inspired to perform a full stop. However, overall, we counted a low number of full stops (14 out of 416). This suggests a rather low impact of the shape, but future studies should avoid this flaw. Further, we did not ask for driving experience — a measure that could have fostered understanding of the data. In addition to that, the eye tracker used in this study could have impaired performance. It was mounted on to the 3D glasses which, in addition to the glasses by itself, reduced comfort. Mounted standalone cameras were not possible because the frame of the 3D glasses obstructed their vision. We also have to mention that, by repeating the take-overs, the drivers had a certain training in performing this very take-over. To mitigate that, we change the location of the hidden vehicle. While habituation and training are recommended for take-over situations [Large et al. 2019], future investigations should investigate various critical situations to confirm our findings. Finally, TOR-Large was slightly larger than TOR-IC and TOR-FoA. With that, we followed recommendations of the pre-test and intended to improve the understanding of TOR-Large, but might have impaired the separation of variables in our design.

Overall, our results indicate that a careful selection of the location where a visual take-over notification is presented, influences take-over performance. Results also suggest that the presentation of a visual map-like take-over notification in S3D can improve take-over performance. Future work should further investigate these aspects, e.g. by exploring various scenarios with different complexities (e.g. more complex visual TORs). It also seems necessary to verify the findings in a motion-based driving simulator or on a test track.

A User-elicited Gesture Set

9

9.1   Problem Statement

> Parts of this section are based on Weidner and Broll [2019b].

Up to now, research on 3D displays in vehicles has focused on information perception and increasing task performance. All previous S3D automotive user interfaces and prototypes (ours and related work) were non-interactive. S3D content could not be manipulated by the user. However, to fully exploit user interfaces that present binocular depth cues, interaction is a fundamental building block. Core objective of this case study is to derive appropriate interaction techniques for 3D content on S3D displays.

We do this in the context of non-driving related tasks (NDRT) — another problem area for future automated vehicles. While previous case studies allowed participants to disengage from the driving task, they also required them to be fallback ready in case of a take-over request. In vehicles with SAE level 4 automation, this is not necessary anymore. While take-over requests are still necessary, for example when the vehicle leaves its operational design domain, they are scheduled and provide the driver an appropriate amount of time to react. Any dangerous and time-critical situation can be handled by the automation system — as long as it operates in its operational design domain. That frees users and enables them to perform a non-driving related tasks. Such task can be, for example, watching a movie or gaming [Jorlöv et al., 2017; Large et al., 2018], but also work-related tasks like

Electronic supplementary material The online version of this chapter (https://doi.org/10.1007/978-3-658-35147-2_9) contains supplementary material, which is available to authorized users.

filing documents, taking part in a training, or reading documents like manuals or schedules [Teleki et al., 2017].

One challenge is that current vehicles are not designed for such activities. Users could put a notebook on their lap and work, rest the phone on the steering wheel, put a tablet on the passenger seat, or just hold and handle pieces of paper. However, today's cars do not necessarily allow for an ergonomic and comfortable way to use these strategies. In return, that can lead to safety and health risks but also in refraining from engagement in NDRTs [Large et al., 2018]. Also, these devices currently do not know about the driving situation and about upcoming scheduled take-over requests [Pfleging et al., 2016]. Hence, information issued by the vehicle might be missed. Another issue is, that in case of a scheduled take-over request, the devices need to be stowed away safely to avoid them interfering with the driving task. It is possible to solve all these issues and enable the use of nomadic devices or traditional pen-and-paper tools in the vehicle. However, another approach is the use the in-car technology for NDRTs. Many cars have already been equipped with displays and touch-screens. It is likely that this trend continues. By utilizing these displays and enable interaction with appropriate input modalities, the execution of non-driving related tasks could be supported. Also, it has been shown that, in certain cases, S3D UIs perform better than 2.5D UIs. For example, investigating tabletops, Bruder et al. [2013] found that the interaction with mid-air touch outperforms 2.5D touch for content that is further away from the screen. Further, Hachet et al. [2011] present a system that combines 2.5D and 3D metaphors and by that, successfully enhances interaction with 3D content. Also, 3D desktops have been shown to outperform 2.5D counterparts in areas like task management, joy of use [Robertson et al., 2000], but also targeting, aligning and pattern-matching [Nacenta et al., 2007]. In addition to that, previous research in cars has shown that S3D can support information structure and user experience [Broy et al., 2015a].

Up to now, our and prior research dealt with 3D content that was non-interactive or purely uni-directional: from the system to the user. Hence, this case study is exploring the interaction of S3D displays for the execution of various NDRTs.

9.2   On NDRTs and In-Car Interaction

Today's in-car interaction is based on few input modalities. Industry heavily relies on buttons and sliders, touch-screens, and joystick-like controllers. Some functions are also available via voice commands. Few cars support gestures on the steering wheel or mid-air gestures. These gestures are mostly used for system-control tasks (e.g. on/off, higher/lower) [Fariman et al., 2016] or menu control tasks [May et al., 2017].

For such gestural interaction, the interaction volume used by drivers is mostly in the center of the car right in front of the center stack [Riener et al., 2013]. Considering gestures and the vehicle, another important factor is that the car is usually moving. Movements can be distorted and then a gesture might not be recognizable by the system. Also, accuracy, e.g. for pointing or targeting in general can be impaired [Ohn-Bar et al., 2012]. Approaches that facilitate gaze tracking [Roider and Gross, 2018] to enhance targeting or use statistical movement prediction [Ahmad et al., 2015] have been developed to mitigate this issue.

Up to now, drivers can use all these interaction techniques mostly to control the car and its features. In a future with automated vehicles, other tasks will be performed and therefore, an in-car user interface that supports such activities is appropriate. According to several surveys, users want to avoid wasting time and spend time productive [Cyganski et al., 2015; Jorlöv et al., 2017]. While driving an automated vehicle, users might prefer tasks like reading or consuming media [Large et al., 2018; Pfleging et al., 2016]. However, results also indicate that participants not necessarily consider passive activities like watching out of the window, listening to music as "wasted time". It is noteworthy that not only well-being and relaxing activities are valid candidates for NDRTs but also work-related tasks [Frison et al., 2019]. While these activities are not mandatory yet, a study by Teleki et al. [2017] found that employers also might require their employees to perform NDRTs while the car is driving automated. For example, time-slot management tasks, preparation of future trips, as well as participating in training and online courses were considered promising activities for people who will spend a large portion of their working hours in an automated vehicle.

9.3   Study Design

For our study, we exemplary selected tasks that can occur during work and relaxing activities (media consumption, reading or document work, controlling applications and documents). In a driving simulator experiment, we want to investigate how users execute such tasks in the restricted car environment.

For this purpose, an elicitation study following the procedure of Wobbrock et al. [2005] was executed. In this type of study, participants see a series of referents and propose interaction techniques. A referent is an action that happens on the user interface. For each action, participants are asked to propose an interaction technique that they would have performed to initiate or execute this action. Vogiatzidakis and Koutsabasis [2018] and Groenewald et al. [2016] provide systematic literature reviews providing an overview of elicitation studies and mid-air gestures. This procedure results in a set of interaction techniques, called symbols. The proposals of participants are then grouped and evaluated. Elicitation studies are often used for deriving and developing interaction techniques for new user interfaces.

Hence, we invited participants to propose gestures for 24 referents. Referents are carefully selected to contain various shapes, sizes, and locations. We calculated agreement and propose a user-defined gesture set for the interaction with S3D content in-cars.

Elicitation studies are prone to bias [Tsandilas, 2018]. Regularization bias describes that users implicitly create a set of rules. Then, they try to base all their proposals on these rules. Harmonic bias is the fact that users try to re-use gestures. Procedural bias is based on the setting in the experiment. Often, elicitation studies are performed with low-fidelity prototypes which makes it harder for participants to imagine the situation. This lack of fidelity in return influences their answers. Finally, legacy bias often influences results. When participants have a hard time thinking of an appropriate interaction technique, they try to apply known interaction techniques [Morris et al., 2014]. Especially for the interaction with S3D content, legacy bias can be a problem due to prior knowledge of WIMP techniques (windows, icons, menus, pointer). Here, participants are likely to propose 2.5D interaction techniques for the S3D interface which might be inappropriate. However, legacy bias can also help to create rather intuitive user interfaces because it introduces known and established techniques into the final set of symbols [Köpsel and Bubalo, 2015].

9.3.1   Procedure

When participants had arrived, they filled out a consent form that also stated the purpose of the study. They then had to pass a stereo vision test [Schwartz and Krantz, 2018]. If they passed, they were asked to take a seat in the simulator and to adjust the seat so that they can reach the dashboard but still sit comfortably. Following, they were handed a tablet to fill out the introductory questionnaire and got an introduction on automated driving. Then, they were required to perform a priming task to mentally prepare them for 3D interaction [Morris et al., 2014]. Here, they were asked to put various shaped wooden blocks in a wooden box (c.f. Figure E.1 in the electronic supplementary material E). After the priming task, they put on gloves and stereo glasses. Both were tracked by a motion tracking system. The gloves were necessary to record hand positions (c.f. Figure E.2 in the electronic supplementary material E).

The experimenter started a video that showed a car drive. With that, the elicitation process started. Participants saw a referent and were asked to provide at least two symbols aka interaction techniques. If they had not indicated their favorite yet, they were asked to do so. The experimenter was located outside the peripheral view behind participants to their left. Participants were repeatedly instructed to think-aloud and that they should ignore technical limitations like possible inaccurate tracking systems. If participants mentioned secondary devices like mouse, rotary controller (c.f. Figure E.3, electronic supplementary material E), or keyboard, they were given to them. This process was repeated for the 24 referents. After the 24th referent, the elicitation process was finished. Participants were asked to fill out the SSQ and the MSSQ, and a set of questions about previous usage of devices.

9.3.2   Apparatus

Simulator
The driving simulator used the design described in Chapter 4. We showed people a video of a drive along a coastline[1] on the front screen to provide participants a sense of driving in an automated vehicle. The simulator had no gear stick but an armrest.

In this SAE level 4 automation system, the car performed the whole dynamic driving task. It never left its operational design domain. The driver does not need to be fallback-ready.

Referents
The referents of the elicitation study are based on the described prior work investigating the preferred activities of users in automated vehicles. These studies agree that media consumption and work-related activities (e.g. document filing, reading) are tasks that are likely to be executed in automated vehicles. To allow for generalization, we also include an abstract scenario with context-free objects. For the final set of referents, we created 4 scenarios: *abstract, two documents, document stack,* and *video player.* The four scenarios contain 24 referents in total.

The following paragraphs describe the referents and scenarios. Numbers in brackets are identifiers (ids) of the referents and are used throughout this section as identifiers.

Abstract task Figure 9.1a shows the layout for the scenario *abstract.* It contains 9 referents that requires participants to select (1), move (2, 3, 4, 5), rotate (6, 7) and scale (8, 9) cubes. For the action 1, 2, 4, 6, and 8, cubes were positioned -5 cm behind the screen and had an edge length of 5 cm. In 3, 7, and 9, the cubes were

---

[1] www.youtube.com/watch?v=whXnYIgT4P0, 2020-02-02

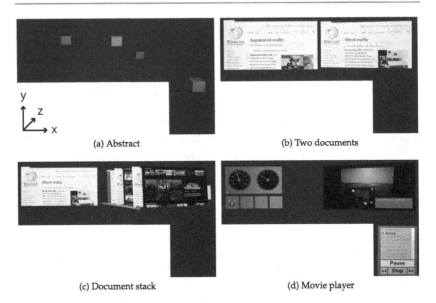

(a) Abstract   (b) Two documents

(c) Document stack   (d) Movie player

Figure 9.1 The four scenarios used in the elicitation study. The illustrations show the state of the UI after all referents have been shown to participant

positioned +5 cm in front of the screen and had an edge length of 10 cm. Cube 4 and 5 inverted their position regarding the projection plane (from +5 cm to -5 cm and vice versa, edge length: 5 cm). That means that they started in front of the screen and then moved behind the screen (and vice versa).

Two documents   Figure 9.1b shows the scenario *two documents*. The left window is 5 cm in front of the screen, the right one 5 cm behind the screen. The five referents of this scenario are rotating the right document (10, 5°) to make it face the user, selecting a hyperlink (11), increasing size of both documents (12, 13) and highlighting text on the right document (14).

Document stack   In Figure 9.1c, the scenario *document stack* is illustrated. The left document is located 5 cm behind the projection plane. The stack's center is 12.5 cm behind the projection plane. It has a total size 25 cm. The five referents of this scenario are as follows: bring the document stack closer (10 cm of the 25 cm deep stack in front of the screen). Then scrolling up and down in the foremost document using a scroll-bar on the right side (16). After that, participants had to scroll through the document stack using a scroll bar on the lower left side of the stack (17). Then rotating the document stack (18, 30° counterclockwise in yaw) and moving the middle document of the stack to the right side (19).

Movie player   The final scenario is illustrated in Figure 9.1d. Similarly to the previous two scenarios, it has five referents. All objects are located 2.5 cm in front of the projection plane. Participants had to move (20), rotate (21), and scale (22) the movie player. In this intermediate configuration, there was an intended overlap between instrument cluster and movie window. Participants were required to move it (23) and then start the video using the control panel (24).

The scenarios were designed to contain referents that represent the canonical tasks *move*, *rotate*, *scale*, and *select* (c.f. LaViola Jr. et al. [2017]). Each scenario contains at least one of the canonical tasks. By having various different referents executing the same canonical task, we intend to derive interaction techniques that are likely to work across specific tasks. The referents were further positioned in such a way that each canonical task had to be performed at least once with an object behind the projection plan, in front of the projection plane, and with an object that appeared in front and behind the projection plane (mixed). The object's positions were chosen by us and were all within the zone of comfort. A list of the single referents with their depth location, and canonical tasks can be found in the electronic supplementary material E, Table E.1. The single referents also cover heterogeneous application domains. Participants experienced the scenarios in random order but not the tasks.

9.3.3   Measures

The initial questionnaire asked participants for age, gender, handedness, and motion sickness susceptibility (MSSQ). Also, how many km per year and how many days per week they drive their car. It also contained a questionnaire asking about the prior usage of devices and technology. The full questionnaire is in the electronic supplementary material E.2.

During the study, participants were encouraged to think-aloud. Further, we made audio and video recordings as well as recordings of the hands' and head's position. The recordings of the hand movement are later used to analyze if the hands had collided with the virtual objects. For that, we used rigid virtual hand models[2] and added them to the UE4 application. Then, we animated them using the recorded tracking data and noted collisions. While this method is not fully accurate due to

---

[2] https://developer.oculus.com/downloads/package/oculus-hand-models/, 2020-09-24

hand scaling and finger movements, it allowed for an estimation on how participants intended to manipulate the referents.

We further calculate the agreement rate (AR) [Vatavu and Wobbrock, 2015] of the set of proposed interaction techniques. Calculating agreement requires classification and grouping of interaction techniques. Originally, the groups were intended to contain only exactly the same gestures. However, Piumsomboon et al. [2013b] argue for a loosened constraint that requires that similar gestures are grouped together for calculation of agreement because small variations could prevent agreement. For example, using the loosened constraint, *point at object with the index finger* and *point at object with index and middle finger* are grouped to *point at object with finger(s)* and do not result in two distinct groups. We further report the chance-corrected agreement κ [Tsandilas, 2018] which considers the fact that participants could propose similar gestures purely by chance. Both measures, AR and κ, describe how much participants' proposed gestures differed. We further calculate co-agreement between scenarios according to Vatavu and Wobbrock [2015], considering canonical tasks, scenarios, and the shape of object (flat or volumetric). After the experiment, participants were required to fill out the SSQ. That concluded the experiment. Approximate experiment duration was 35 minutes. Participants had the chance to win 50 EUR.

9.3.4 Sample

Overall, 23 participants took part in the study (12 male, 11 female). They were invited using convenience sampling (university mailing list, Facebook groups, and direct contact). They were mostly staff and students of a local technical university. Their age ranged from 21 to 59 with a mean age of 31.00 years (SD = 10.96).

On average, they drive their car on 3.7 (SD = 1.89) days per week for a total of 6.56 (SD = 6.67) hours. Figure E.4 in the electronic supplementary material E depicts their annual mileage. The majority of participants (more than 21) uses multi-touch devices and on-screen gestures often. More than 17 rarely or have never used augmented reality, virtual reality, or 3D mid-air gestures. Around 50% use 3D displays sometimes or often. Similarly, around 50% sometimes or often play games on their desktop pc or smartphone. Figure E.5 in the electronic supplementary material E illustrates the previous usage of devices.

Participants were instructed to adjust the seat to be comfortable. 19 participants could touch the whole dashboard without leaning forward. Mean MSSQ score was M = 10.4 (SD = 7.72). There were no dropouts.

9.4   Results

The simulator sickness scores of participants had a total score of TS = 6.48
(SD = 7.13). The single sub-scales were oculomotor O = 3.30 (SD = 3.11), disorientation D = 6.48 (SD = 7.13), and nausea N = 1.13 (SD = 1.69). Hence, we can
assume that results were hardly negatively influenced by symptoms of simulator
sickness.

9.4.1   Interaction Techniques

We recorded 1104 interaction techniques (23 participants × 2 options × 24 referents). We had asked participants for their favorite, thus the final set consists of 552
symbols. 36 out of 552 or 6.52% were not gestures. The 36 included eye-tracking,
brain-computer interfaces, voice commands, and traditional input devices like buttons or controller.

The 552 gestures were classified according to the taxonomy for surface computing [Wobbrock et al., 2009] and the extension for augmented reality [Piumsomboon
et al., 2013a]. Most of the gestures belong to the group *static pose and path* (53%).
Most required *continuous* feedback from the user interface (81%) . *Physical* gestures (74%) dominated, indicating that participants perceived the objects as tangible.
Because of that, most gestures operated in the *object-centered* coordinate system
(68%). 73% were executed using only the dominant hand (*unimanual dominant*)
and 70% were performed *in-air*. Figure E.6 in the electronic supplementary material
E shows an overview of the classification.

In an augmented reality application, users can interact in several ways with the
virtual content. They can try to manipulate it in a *direct* way, e.g. by trying to touch
the virtual object. This only works if the object can be touched and is not out of reach
or blocked as in our case: objects with positive parallax are blocked by the physical
screen. Here, they can try to manipulate it directly but are blocked by the screen,
hence they can only manipulate it in an *indirect* way, e.g. by trying to reach but
when the screen blocks their movement, they perform a gesture on the screen. They
can try to manipulate it via a *proxy* object, e.g. by manipulating an imagined object
somewhere but not at the location of the virtual object. Finally, they can execute
the manipulation with a third-party device, ignoring the virtual content and by that,
execute a *non-gestural* interaction. To get a better understanding of our gesture
set, we further extend this taxonomy with the category *manipulation strategy*. By
using the dimension *direct*, *indirect*, *proxy*, and *non-gestural*, we can model the
intention of the user and better understand how the proposed symbol relates to the

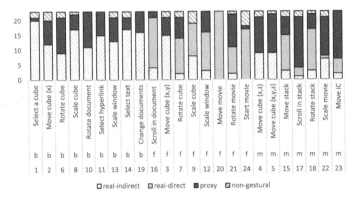

Figure 9.2 Results describing the manipulation strategy. Objects behind the screen were often manipulated real-indirect. Objects in front of the screen often real-direct or by proxy

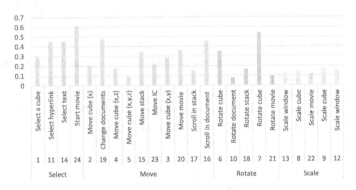

Figure 9.3 Agreement rate per referent grouped by canonical task. Scale received rather low agreement rates, select overall high agreement rates (0 – 0.1: low agreement, 0.1 – 0.3: medium agreement, 0.3 – 0.6: high agreement, ≥0.6: very high agreement)

virtual objects. Figure 9.2 shows the results of this analysis. Overall, there were 204 real-indirect, 151 real-direct, 161 proxy, and 36 non-gestural interactions.

Especially objects in front of the screen — labeled with f — are primarily manipulated directly or using a proxy object. Many of the referents behind the screen were executed using proxy and real-indirect interaction techniques. Here, participants either operated on an imagined object or tried to manipulate the virtual object directly but were blocked by the screen. The latter could lead to an uncomfortable experience in real-world applications. Only the classification for task 3, *Move cube (x,y)* (moving a cube in front of the dashboard, parallel to the dashboard) does not follow this scheme. Here, real-indirect gestures dominate. Both, 1 and 2, resulted in a similar pattern and participants often proposed touch-screen inspired gestures. It is likely that the mental models build by these tasks carried over and influenced results for task 3. In task 4, which required participants to propose a gesture for an object moving away from them, touch-screen like gestures were less appropriate. Hence, most participants proposed proxy gestures for this task. Overall, results suggest that participants perceived the objects as spatial and tangible.

Table 9.1 Co-agreement rates per canonical task and referent groups. Only few received medium co-agreement (bold). Most received low or no agreement

Group	CAR Move	n	CAR Rotate	n	CAR Scale	n	CAR Select	n
All	.000	n=10	.000	n=5	.000	n=5	.040	n=4
Scenarios with cubes	.004	n=4	**.233**	n=2	.032	n=2	**.300**	n=1
Scenarios with use cases	.000	n=7	.000	n=3	.004	n=3	.099	n=3
Flat	.087	n=4	.016	n=2	.004	n=3	**.217**	n=2
Volumetric	.000	n=6	.051	n=3	.004	n=2	**.186**	n=2

9.4.2 Agreement

Grouping the 552 gestures into classes of similar symbols resulted in 68 classes. 29 for move operations, 19 for rotation, 15 for scale, and 5 for select. Using these classes, we calculated agreement rate AR. The overall agreement rate according to Vatavu and Wobbrock [2015] is $AR = 0.271$ $(SD = 0.159)$. That results in a disagreement rate of $DR = 0.728$ $(SD = 0.159)$. This agreement rate is usually calculated and used to provide an estimate on how aligned participants proposed symbols were. However, Tsandilas [2018] provides a critical discussion of agreement rates and their proneness to being based on pure chance. Based on that, they derive the chance-corrected agreement rate κ. For our elicited gestures, the chance-corrected agreement rate is $\kappa = 0.232$. Both, κ and AR stand for medium agreement [Vatavu andWobbrock, 2015].

Figure 9.3 shows the agreement rates per referent grouped by the canonical tasks (electronic supplementary material E, Figure E.7) shows the agreement rate sorted by task). Agreement of the selection task ranges from 0.3 to 0.61 and by that consistently shows medium to high agreement. Agreement of the canonical tasks *move* and *rotate* varies widely. For the *scale* operation, agreement is relatively low.

9.4.3   Co-agreement

Table 9.1 shows the agreement rates for various combinations of groups. Co-agreement describes the agreement of groups of symbols to each other. We analyzed agreement over the canonical tasks as well as the scenarios and the shape of the elements. Most of the combinations achieve *no agreement* ($= 0.000$) or *low agreement* (≤ 0.100). Few combinations achieve medium agreement ($0.100 - 0.300$), namely select and the abstract rotation.

9.4.4   The User-Defined Gesture Set

With the agreement rates for the single tasks, we can evaluate how similar the proposed gestures were. The fewer distinct gestures per task, the higher the agreement. By that, it gives an estimate on how intuitive the most-mentioned gesture for a task is. By selecting the gestures that achieved the highest agreement rate AR per single referent, the final set of gesture contains 12 distinct gestures. They are shown in Figure 9.4. Colors indicate type of canonical tasks. The letters are chosen to confirm with the overview provided in the electronic supplementary material E, Table E.2. All, except 3, are mid-air gestures (B,I, and K). Those 3 are traditional touch-screen gestures whereas one of them (I) relies on touch duration to indicate the angle of rotation.

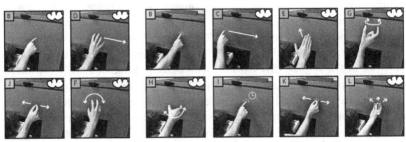

(a) Best agreement rate per   (b) Best agreement rate per referent.
canonical task

Figure 9.4 Core gesture set (9.4a) with extension (9.4b). Colors indicate canonical tasks (red = rotate, green = select, yellow = scale, blue = translate). Gestures with a cloud pictogram (all except B, C, and I) are mid-air gestures and not touch-screen gestures

Figure 9.4a shows the gestures that achieved the overall highest agreement rates per canonical task. Results suggest that these gestures are most intuitive across our sample. Figure 9.4b shows the additional gestures that resulted in high agreement. These eight gestures can be considered as an extension or alternatives to the four best gestures.

9.5   Discussion

In this study, we wanted to investigate how users interact with S3D content in vehicles. Our final set contains twelve gestures. The resulting overall agreement is $AR = 0.271$ (SD = 0.159). Chance-corrected agreement is $\kappa = 0.232$ (SD = 0.159). Co-agreement was low in most and medium in a few cases. To get our final gesture set, we calculated agreement rate for all referents. The gestures with the highest agreement rates were then chosen to be in the final set. Overall, twelve gestures were included: 2 for selection, 3 for translation, 3 for scaling, and 4 for rotation.

9.5.1   Gesture Set

In the end, we decided to split up these twelve gestures into two distinct sets, a core set, and an extension. That offers several advantages. For an application that is built to be used by many people and a wide user base, using the core set with extension promises a user interface that is very intuitive. By providing several gestures for the same canonical task, it is more likely that the gesture a user favors, is included in the set. For prototypes, case studies, and other applications (e.g. for pre-studies), only the basic set can be used. It offers gestures that showed the highest agreement rate and which are likely to be used by participants. Because the set was derived using a variety of tasks, shapes, and scenarios, it is also likely that the gesture set translates to other scenario as well.

A comparison with previous sets might further shed light on the appropriateness and intuitiveness of our set. In this context, it is noteworthy that the derived gestures set, while being develop for in-car interaction, has some similarities to previously developed gestures sets from other domains. Hence, we compare our set with sets that have been derived in studies that focused on the interaction with augmented reality content while wearing an HMD (G1, [Piumsomboon et al., 2013b], AR = .295), on the manipulation of CAD content in a work environment (G3, also while wearing an HMD, [Pham et al., 2018], AR = .317), and on the manipulation of 3D objects on a monoscopic display approximately 3m away (G4, [Khan and Tunçer, 2019],

Table 9.2 Comparison of elicited gestures with other sets G1 – G6. Numbers in brackets indicate total number of gestures for a task from the referenced gesture set. Letters indicate which gesture of our set is also in the referenced one

	Tasks	Device	Objects 2.5D or 3D	Visualization S3D or 2.5D?	Canonical tasks			Total	
					Select	Scale	Rotate	Move	
G1	Abstract tasks; random object	AR HMD	3D	S3D	A (1)	J (3)	G, H (3)	D (2)	5/9
G2	Abstract tasks; random object	Tabletop	2.5D	2.5D	B (2)	K, L (4)	– (1)	C (2)	4/9
G3	Architecture	AR HMD	3D	S3D	– (2)	– (2)	F (3)	D (2)	2/9
G4	3D CAD tasks	50" screen	3D	2.5D	– (0)	– (2)	F (3)	– (2)	1/7
G5	Menu control; center stack	6.4" × 4" screen	2.5D	2.5D	No similar gestures (no canonical tasks)				0/6
G6	No specific tasks; system control	No visual output	–	–	No similar gestures (no canonical tasks)				0/12

no AR mentioned). We also include a gesture set that has been derived while using a tabletop with monoscopic display (G2, [Wobbrock et al., 2009], AR = .30). In the automotive domain, two gesture sets have been defined using similar procedures which we also include in our comparison. First, a set to control in-car menus [May et al., 2017] (G5) and second, a set that has been derived to perform system control tasks (e.g. control the radio, the A/C and general in-car functions, G6, [Hessam et al., 2017]). Table 9.2 list all these gesture sets. Comparing our results with these studies puts the gestures in context and provides suggestions for future development and applications.

Comparison is performed based on the canonical tasks performed in our study and in the cited studies. For the automotive domain, no such studies were available to this time. Thus, we use two studies that have derived gestures for system control tasks. Considering the results, some similarities are noticeable.

From our core gesture set, that contains the four *best* gestures, 3 are also in the final gesture set of Piumsomboon et al. [2013a] (G1). That can be taken as an indicator of the intuitiveness of these gestures across tasks and domains. Only the rotate gesture of our basic set was not in theirs. However, two gestures from our extension are (G, H). This study was rather similar to our ours, except participants were seated in front of a desk and not in a car. Participants saw a virtual object in front of them and were asked to manipulate it. Most differences are based on the fact that their set contains many bimanual gestures whereas ours contains only unimanual symbols. Sets 2 – 4 in Table 9.2 also show overlap with our basic set (B,D,F) as well as the extension (C, K, and L). Main reasons for the smaller overlap with the core set is probably that in the environment of Pham et al. [2018] (G3), some objects were located far away and life-sized (e.g. a couch or a TV) which again, led to different gestures. Similarly, in the setting of Khan and Tunçer [2019] (G4), participants were not able to touch the objects because the screen was 3 m away. In both cases, the environment in which AR was applied impacted the likelihood of any overlap in the resulting gesture sets. That is because the car environment and interaction space inside the car does not foster the usage of rather large objects (e.g. couch) but only of relatively small objects. Another issue, that is special for the in-car environment compared to the settings in studies 1 – 4, is that the environment heavily limits the possibilities for gestures. Many of the gestures in the related gesture sets were bimanual or executed with the left hand. However, in a vehicle with a traditional layout, the steering wheel makes bimanual gestures rather uncomfortable (the driver has to reach around or above the wheel) and participants are used to control the car with the right hand (which is closer to the center stack). We believe that this is the reason why our set only contains unimanual gestures. Nevertheless, all of our derived gestures, except E, are in at least one set which suggests that the gestures have a rather decent level of intuitiveness. In summary, the comparison emphasizes that many-to-many mappings are necessary to account to different affordances. That is in line with previous results of Silpasuwanchai and Ren [2015] that showed that many-to-many mappings enhance UX.

The disparity value of an object did not have a major influence on gestures. Most of gestures in the final gesture set are mid-air gestures, regardless of positive or negative disparity. These gestures target either the actual virtual object (negative disparity) or a proxy object (positive disparity). Nevertheless, three gestures were touch-screen gestures: select by touch (B), rotate by touch (I), and scale by pinch (K). We assume that these gestures are partial artifacts of legacy bias. All three have strong relationships to traditional and known interaction techniques that can be executed on tablet PCs or touchscreens.

Overall, we provide a gesture set for the interaction with S3D content in automated vehicles. The set has similarities to previous gesture sets from other domains which shows the intuitiveness of it. It also has distinct features that emphasize the necessity of domain-specific investigations. With the basic set and its extension, it provides the foundation for prototyping user 3D UIs in vehicles.

9.5.2   Limitations & Future Work

While we ensured that participants were able to reach the dashboard, the additional restriction introduced by a seat-belt was not present in our setup. Hence, we cannot guarantee that this additional freedom of movement did not slightly influence our results. Also, participants wore gloves with the tracking targets and other interaction devices were absent (they were only added if participants mentioned them). On the one hand, this might have helped to reduce legacy bias and propose appropriate symbols. On the other hand, it might have biased people to propose gestures instead of other symbols. Further, we noticed during analysis that answers from each participant did not vary widely, despite them being required to produce more than one symbol and chose their favorite. Hence, the final set is likely to be influenced by regularization and harmonic bias. In this study, we focused on object manipulation that might arise during the execution of work- and well-being related tasks. We deliberately excluded typing as a different problem domain. Future investigations could and should combine our approach with research that has been dedicated to typing while driving [Schartmüller et al., 2018]. Similar to that, research that enhances gesture execution with feedback (e.g. Shakeri et al. [2018] propose mid-air ultrasonic feedback) could provide a viable future research direction.

The derived gesture is designed to be used for various well-being and work-related activities (e.g. gaming in semi-automated vehicles [Lakier et al., 2019] or the mobile office [Janssen et al., 2019; Kun et al.,2016]). In the future, it needs to be evaluated in a real car or at least in a simulator with a moving base. While a preliminary evaluation in a fixed-base simulator might provide additional insights on user experience, the lack of movement that influences gesture execution and performance makes it necessary to evaluate them with additional movement. Such a study should, at best, include longer drives to also investigate physical exertion (e.g. via PACES - Physical Activity Enjoyment Scale [Kendzierski and DeCarlo, 2016]).

In summary, we propose a user-elicited gesture set for the execution of canonical tasks in vehicles equipped with S3D displays. We considered aspects of location, disparity, type of task, and shape of object during development of this gesture set. In addition to that, a comparison with other sets, designed for the manipulation of virtual content in general, showed partial overlap with our set. That substantiates our set's intuitiveness.

Part IV

Summary & Outlook

Discussion: On the Potential of S3D Dashboards

10

Prior work has investigated small-scale S3D displays in cars. Mostly, the instrument cluster was replaced by such displays. Results of these studies were encouraging to venture further in the domain of automotive S3D applications. Intention of this thesis was to do exactly that and broaden the knowledge about the potential of S3D displays in cars. Special attention has been paid to the fact that the available display space in cars can be large and by that, can provide new opportunities for designers and developers. The objective was to explore if S3D dashboards can help solving some of todays and future problems in manual and (semi-)automated driving. Based on this motivation, three research questions were formulated (c.f. Chapter 1):

RQ1: How can large interactive stereoscopic 3D dashboards be designed and developed?

RQ2: What are viable applications for large interactive stereoscopic 3D dashboards?

RQ3: What are appropriate interaction techniques for stereoscopic 3D dashboards?

The process of answering these questions consisted of three major steps: First, we identified potential application domains and open problems in the automotive domain. Second, we developed a prototyping environment to test applications that use a large S3D dashboard. Third, in several driving simulator studies, we explored the potential of S3D to tackle prominent issues. This chapter recapitulates and discusses the presented results in the context of the research questions and provides an extrapolation of the findings to future use cases.

F. Weidner, *S3D Dashboard*, https://doi.org/10.1007/978-3-658-35147-2_10

10.1   Design and Development of S3D Dashboards

RQ1: How can large interactive stereoscopic 3D dashboards be designed and developed?

To answer the first research question, we created a platform for rapid prototyping of automotive S3D user interfaces (Chapter 4). An initial selection of criteria that are essential for the rapid prototyping and evaluation of S3D dashboard applications was used for a comparison with existing development tools.

Based on this comparison, it was deemed appropriate to create a custom platform to build S3D dashboard prototypes. The platform consists of a prototyping software based on Unreal Engine 4 and a mock-up of a car. The software has two parts: an environment simulation and the UI prototyping toolkit. Both are connected via network. The mock-up of the car provides a flexible simulation of a car interior. To realize S3D, rear-projection and a head-coupled perspective via head-tracking has been applied. With these tools, various studies, including numerous in- and output devices, have been executed. Through iterative refinement of the platform — e.g. integrating head-tracking, edge-blending, and improving the mount of the steering wheel — reasonable simulator sickness scores have been achieved and visible display space has been maximized. Using this platform, we further derived a zone for comfortable viewing that constrains the effective range of disparity values (Chapter 5). It has proven effective in several user studies and guides design of S3D UIs in vehicles. The platform provides flexibility and reconfiguration in hardware and software. In addition to steering wheel and pedals, various in- and output devices have been connected. For example, a keyboard or an external software tool for a wizard-of-oz study. The decision to build the platform and not to reuse an existing tool was mostly driven by the requirements and availability. While it introduced additional workload, it provided immense flexibility. Other development teams later also decided to build their platforms using similar approaches. Their choice can be seen as validation of our approach. Using the game engine Unreal Engine 4, we have been relying on up-to-date tools and methods as well as visualization techniques. However, the short release cycles have made it necessary to regularly update prototypes to benefit from these technologies. By that, backward comparability was not always provided. It also required repeated and exhaustive tuning of the vehicle's physics model. Furthermore, the design of the buck including rear-projection — while working — has two core limitations. First, despite the buck having wheels and handle-bars to adjust parts like the seat and wheel, the weight of the single parts makes moving them a rather strenuous task. A lighter build or a smoother mechanism would solve this issue without impairing stability. Second, the interplay of projector and projection screen is crucial for appropriate user interfaces. Especially the quality of the projection screen is important. It needs to provide uniform gain in brightness and color across the whole screen [Herrick, 1968]. In our study, we had to carefully adjust colors and brightness of the UI elements to mitigate these issues. The driving simulator has only a fixed base and is not mounted on a moving platform. As mentioned in Section 2.3, moving-based simulators exist but are not mandatory for research on automotive user interfaces. Despite that, having no motion cues is still a factor that needs to be considered when interpreting the results of all our user studies.

Overall, the system allowed for rapid prototyping of various S3D user interfaces that included motion tracking, voice interaction, wizard-of-oz studies, and traditional input via buttons. The flexibility of our environment, coupled with the open-source nature and the rather lightweight design offers an environment for future exploration of S3D user interfaces. The derived zone of comfort is similar to previous results but also shows certain peculiarities. Especially the rear-projection mode needs to be considered for prototypes as it introduces areas with a lower effective depth budget. The final volume can guide design and helps to avoid the pitfall of excessive disparity. In the future, the simulator environment should be coupled to a validated off-the-shelf environment simulation solution. To enhance user experience and to explore the full potential of S3D dashboards, glasses-free solutions should be explored, e.g. by integrating lightfield or large autostereoscopic displays into the setup.

10.2 Applications for S3D Dashboards

RQ2: What are viable applications for large interactive stereoscopic 3D dashboards?
Considering current and future use cases, we wanted to assess if and how S3D dashboards can contribute to solve them. By that, we want to not only explore current applications, but also provide an informed outlook on future applications and prospective research avenues. To do this, we analyzed literature to find current and future challenges in the automotive domain. Based on the results of this analysis, which was based on survey papers and workshop results, we selected three use-cases:

1. Driver distraction (Chapter 6)
2. Trust in Automation (Chapter 7)
3. Take-over Requests (Chapter 8)

For each use-case, a user study was conducted to see if and how S3D can be a viable solution. In the next paragraphs, we will discuss the findings and summarize them in the context of the research question.

10.2.1 Driver Distraction

In the first use case, *driver distraction*, we wanted to investigate the potential of pure binocular depth cues for information visualization across the dashboard and if the application of such a cue, can increase performance in change detection and list selection. By that, we hoped to mitigate driver distraction posed by these tasks during manual driving.

In a driving simulator study we operationalized these two tasks in form of abstract versions of real-world examples. Change detection was realized by participants being required to quickly react to a visual stimulus appearing on the dashboard. During list selection participants had to select a target string on menus appearing across the dashboard as quickly as possible. In both cases, the visual stimulus was either change in relative size or a binocular depth cue.

We found no significant differences in perception of changes and no significant differences in list selection performance. At the same time, manual driving performance in the car following task did not differ. This indicates that this application domain seems not suited for binocular depth cues and large S3D dashboards. The fact that S3D did not result in performance increase in this study on driver distraction opens up an important discussion point regarding the application of S3D for non-spatial tasks. Previous work did measure enhancements in similar tasks ([Broy et al., 2013] important info closer to the user, [Szczerba and Hersberger, 2014], change detection, [Broy et al., 2015a] structuring of information). However, our results indicate that the effectiveness of S3D compared to 2.5D at least relies on other, unexplored factors. On the one hand, it could be that unintentional bias led to experimental designs that favored S3D or to designs that are ill-suited for 2.5D. On the other hand, it could be that certain design patterns simply lead to a distinct benefit of S3D. For example, color has been shown to have an influence [Broy et al., 2015a], but many other factors like spatio-temporal resolution or texture are known to influence binocular fusion and by that, task performance [Howard and Rogers, 2008]. Future studies need to further explore the design space of S3D to justify the application of S3D as a pure design element in vehicles.

In addition to the design of S3D UIs, other factors seem to influence the effectiveness of S3D in tasks that require fast interpretation of and reaction to a visualization. The study designs of the first study (driver distraction) and of the third study (take-over request) have certain high-level similarities. Both tasks required the user to react to a stimulus within a limited time-frame. In the simple environment with a simple and rather abstract task of study 1, no difference between S3D and 2.5D could be detected. However, in the more complex environment with added spatial attributes of study 3 (take-over), S3D outperformed 2.5D. Combined with previous results, that suggest that S3D is beneficial for navigational cues [Broy et al., 2015b], understanding of maps [De Paolis and Mongelli, 2014], and route memorization [Dixon et al., 2009], our results make a case for the application of S3D displays in cars to present spatial and navigational information with binocular depth cues. It is important to note that McIntire et al. [2014] (c.f. Figure 3.1) report that S3D is less effective for navigation tasks. However, they state that in studies where it did not lead to benefits, S3D depth cues were either outside the effective range (c.f. Figure 2.2) or other depth cues, especially due to self-motion in the virtual environment, made S3D less effective. Both drawbacks do not apply to S3D dashboards which present content within the effective range of binocular depth cues and the user is not completely immersed in the virtual environment. Hence, S3D could improve applications like navigational systems that continually support drivers by showing the current route on a map or S3D maps that show information about the surroundings (e.g. sights, service stations, restaurants, or shopping centers). Here, research suggests that S3D could have the potential for reducing driver distraction by increasing understanding of the spatial information.

In Chapter 6, we intentionally refrained from creating a complex UI but to investigate the influence of isolated binocular depth cues (similar to Szczerba and Hersberger [2014] or Pitts et al. [2015]). At the same time, this limits results due to the absence of other design elements, e.g. other depth-cues and texture. However, this way, results can act as a reference for further exploration of the interplay between various design elements.

In summary, reducing driver distraction is a safety-relevant topic. Supporting the use of in-car menus and the perception of changes on the dashboard can reduce driver distraction. Previous research suggested that highlighting important information using S3D is an appropriate tool to do just that. By analyzing the effect of isolated binocular depth cues, we increase knowledge on the effectiveness of S3D to mitigate driver distraction: simply highlighting content by S3D on the premise that important content should be closer to the user and thus, better recognizable, showed not to be universally true. This opens up the research area on what contributes to an effective S3D design, actually increases secondary task performance, and reduces driver distraction. A further investigation on the interplay of design factors is necessary to pinpoint and categorize specific factors that enable performance enhancements by S3D displays. However, using S3D for presenting spatial information — e.g.

for improving the quick understanding of navigational cues, seems promising. By researching S3D design guidelines and evaluating navigational systems that present info in S3D, drivers could be less distracted and enjoy a safer drive.

10.2.2 S3D For Trust

Trust is an important factor when using automated cars. We hypothesized that an S3D agent that acts as a visual representation of the automation system, increases trust.

In a driving simulator study (Chapter 7), the automation system was represented by an anthropomorphized agent. While driving automated, participants could either trust the agent or take-over control and by that, display distrust. The S3D agent's design properties were carefully calibrated and received a slightly positive rating among participants on all scales. By applying the agent, neither self-reported nor behavioral trust increased. However, anthropomorphism increased with presentation in S3D. In addition to that, we could show that anthropomorphism and self-reported trust are positively correlated. Hence, in the context of the research question, S3D was not helpful in increasing trust in automation. Reasons for this are most likely the initial trust building and the limited capabilities of the agent. Nevertheless, a comparison with previous results indicates increased behavioral trust when using an avatar. Also, the correlation of anthropomorphism and trust is a motivating result. It suggests that trust and anthropomorphism are related — if one is larger, the other one is larger as well (and vice versa). While it does not imply a causation, the correlation is worth investigating. Considering that the S3D visualization increased anthropomorphism, we believe that further refinement of the agent regarding design and capabilities can lead to more trust in the automation. In addition to that, our agent's design, and the design decisions it is based on, can be used as a baseline for future research on anthropomorphized agents.

Another application area where S3D could foster trust is by communicating system behavior. Previous research indicates that communicating external awareness of automated vehicles, e.g. regarding pedestrians or upcoming events or actions, can have a positive influence on trust [Koo et al., 2015; Rezvani et al., 2016; von Sawitzky et al., 2019; Wintersberger et al., 2019]. Considering the evidence for benefits of S3D in communicating spatial information (our results, but also e.g. Dettmann and Bullinger [2019]), binocular depth cues could help to communicate the automated car's understanding and interpretation of the environment. By displaying the vehicles current representation of the environment in S3D, the users' understanding could be fostered and by that, trust in the automation system could be enhanced.

It is important to consider that especially the short duration of the drive limits the results of our user study on trust. We chose this duration to enable comparison with previous results. In the future, longer drives that initially build a trustful relationship between human and the automation system seems necessary to further evaluate such technologies and their potential to increase trust.

In summary, an appropriate level of trust in automation is considered fundamental for the adoption of automated vehicles. Based on the comparison with previous results, an ongoing presentation of an agent can potentially be helpful in fostering trust. However, we did not uncover a direct effect of S3D on trust. Nevertheless, S3D increased anthropomorphism which in return had a positive correlation with self-reported trust. In addition to that, considering the benefits of S3D to communicate spatial information, this potential can likely be facilitated to enhance trust by communicating external awareness of an automated vehicle in form of such spatial visualizations. By exploring both aspects, highly anthropomorphic S3D agents and awareness-communicating UIs, a significant increase in trust might be possible.

10.2.3 Take-Over Notification

Take-overs maneuvers are an immense challenge for humans. When the car is driving automated, drivers are susceptible to lose situation awareness which they need to regain quickly when a take-over notification is issued. We assumed that supporting the driver with a take-over notification that displays spatial information will support the take-over process and will improve performance. We further assumed that communicating such cues in S3D can enhance performance even more. In a driving simulator study (Chapter 8), participants experienced the take-over notifications (incl. baseline) which appeared on several locations on the dashboard (instrument cluster, upper vertical center stack, focus of attention).

Results suggest a benefit of S3D compared to 2.5D. Further, the S3D TORs displayed on the instrument cluster and the focus of attention performed better than baseline. Also, there were no impairments of workload, regardless of TOR. Thus, presentation in S3D can positively influence situation awareness during take-over. Considering the relatively small effect of S3D, it is important to cautiously design and test the UIs. For example, simply displaying spatial information in S3D does not automatically lead to improvements as the case of TOR-Large showed: it did not lead to significantly more safe-take over maneuvers compared to baseline. A careful placement of the TOR — in our case the instrument cluster or the focus of attention — and a thoughtful design with respect to information density and layout is necessary, to benefit from S3D. Also, the spatial nature of the visualization that shows the surroundings seems to be an important factor. Notifications conveyed by non-spatial icons are likely to not improve performance as the comparison with the baseline-S3D icon but also the results of the study on reducing driver distraction (c.f. Chapter 6) show. It is important to note that in this study, participants had no reason to not trust the system. Repeating the critical situation ensured that the system was working correctly. However, in a real-life system, it is possible that a non-sufficient trust level leads to drivers ignoring the warning and the information provided by it. For example, if the information provided was wrong once, users might refrain from relying on such systems in the future. Hence, maintaining an appropriate level of trust is necessary to make full use of smart take-over notifications and to benefit from the spatial S3D presentation of the acquired information.

When trying to generalize our results to other scenarios, we are conservative. Extrapolation is limited, especially because we have only tested one scenario. For other scenarios, different visualizations are necessary. That requires a deeper understanding on how much information such an icon can convey before performance degrades. Also, having performed the study in a fixed-base simulator and having evidence that movements of a vehicle influence the take-over decision, this limits our results and should be considered in future studies.

Considering that especially in Level 3 automation, people are prone to slipping out-of-the-loop, it is important to provide safe assistance systems to users and increase safety [Endsley, 2017; Endsley and Kiris, 1995]. Increasing situation assessment during the take-over process seems to be a viable approach. Using simple visualizations that show the surroundings, large S3D dashboards showed potential to do that, to support drivers, and to improve safety during take-over maneuvers.

10.3   Interaction Techniques for S3D Dashboards

RQ3: What are appropriate interaction techniques for stereoscopic 3D dashboards? We investigated the last research question in Chapter 9. There, we presented interaction techniques for S3D content and S3D dashboards in cars. All prior applications that used S3D displays in cars were either non-interactive or used 2D input techniques for the manipulation of 3D content. However, certain future applications, e.g. the interaction with S3D navigation systems, might require more appropriate 3D user interfaces. For this user study, we derived four scenarios — one being abstract, and three representing work and well-being related tasks. Participants then had to propose interaction techniques for 24 referents. These techniques were then condensed in a set of gestures having a maximized agreement rate.

The final gesture set has two parts: a core set and an extension. The core set contains one gesture for each canonical task (translate, rotate, scale, and select). The extension contains additional (redundant) gestures. Having such redundant gestures increases the likelihood of a participant performing a gesture known to the system. These interaction techniques can be used as a foundation for the development and evaluation of interactive applications that present content with binocular depth cues in cars. It is important to note that the interior of a car has distinct properties that are reflected in this set. For example, it does not contain bi-manual gestures which are rather uncomfortable to execute in cars. Nevertheless, the gestures show partial overlap with other sets originating from various domains. That substantiates the intuitiveness of our gestures. While we acknowledge that users probably will not use gestures all the time for all task in vehicles — simply out of convenience but also because of social factors — gestural interaction can be a vital and valid part of multimodal user interfaces. Also, at least for the interaction with S3D content, gestural interaction was the most intuitive modality, considering the few proposed symbols that were not gestures. Because the set is based on various tasks that incorporated various object sizes, contexts, and object locations, a rather high level of generalizability can be assumed.

The focus on canonical tasks was inspired by previous research on 3D user interfaces [LaViola Jr. et al., 2017]. By that, it did not include other use-cases, e.g. typing approaches, which should be done in the future as research suggests that office tasks are viable non-driving related tasks in SAE level 4 automated vehicles. Another important aspect is that in real vehicles, the execution of gestures might be hindered by movements, vibrations, as well as exertion. Hence, future research should investigate aspects like the development of gesture accuracy over time but also the performance of gesture execution while driving in a moving vehicle. In addition to that, considering the overlap with other sets but also because many of the gestures operate on proxy elements, the gesture set could be valid for the interaction with content displayed on other 3D displays, e.g. AR-HUDs.

Conclusion & Future Work 11

This thesis explored the design space and possibilities of stereoscopic 3D dashboards. Important problems that have been and will be a challenge in the automotive domain, guided the research process. Our results provide insights in these very domains — driver distraction, trust, and take-over process — but also open up new application domains.

To investigate and apply stereoscopic 3D dashboards, we developed an integrated prototyping platform. It includes a driving simulation environment, a toolkit for building driving scenarios, a user interface simulation toolkit, and a physical car mock-up that is equipped with an S3D dashboard. Here, an active-stereo projector realizes S3D visualization via rear-projection. For this very environment, we derived a zone of comfortable viewing that allows for the design of 3D user interfaces with content that does not exceed binocular fusion limits (D[$-0.8°$; $+0.58°$]). Using this prototype, we conducted four user studies. We investigated if an S3D dashboard can help to support system control tasks and change detection to reduce driver distraction. No benefits but also no impairments of binocular depth cues on driving performance and secondary task performance could be found. Further, we explored if and how virtual agents that represent the automation system can help to foster trust between the human and the vehicle. While no difference in trust between S3D and 2.5D could be shown, adding binocular depth cues increased anthropomorphism of the agent. Further, the agent led to higher behavioral trust scores compared to previous techniques. In addition to that, we could show a positive correlation between anthropomorphism and self-reported trust. Our results further indicate that S3D dashboards can support take-overs by presenting a visualization that shows the current traffic situation on the instrument cluster or the focus of attention. It is important to note that neither gaze behavior nor workload was negatively affected by our visualization. Finally, as previous applications have been non-interactive, we presented the first gesture set for the interaction with S3D content in cars. Partici-

© The Author(s), under exclusive license to Springer Fachmedien Wiesbaden GmbH, part of Springer Nature 2021
F. Weidner, *S3D Dashboard*,
https://doi.org/10.1007/978-3-658-35147-2_11

pants' responses showed medium agreement with an agreement rate of AR = 0.271. Resulting from our user elicitation study, 12 gestures — a core set of 4 and an extension of 8 — have been derived to perform rotation, selection, scaling, and translation tasks on S3D displays. The extension set contains redundant gestures to increase the likelihood that a user chooses a gesture in that is actually in the set. Together with the medium agreement, our gesture set promises to allow for an intuitive interaction with S3D content in cars.

Arguably, the need or significance of S3D dashboards in the automotive domain is worth discussing. However, previous benefits of S3D outside the automotive domain and especially the encouraging findings of this thesis and the work it is based on, suggest that a further exploration is worthwhile. Our results can be a starting point to make take-over maneuvers safer by supporting drivers during situation assessment. These situations are considered dangerous as the human is out-of-the-loop and might not have situation awareness — an S3D visualization directly mitigates this issue. A positive effect on trust of our S3D agent has failed to materialize. Nevertheless, we could show that the presentation of virtual humanoid agents in S3D can increase anthropomorphism and that anthropomorphism correlates with self-reported trust. This indicates that a more anthropomorphic agent can potentially increase trust in automation — a crucial factor for the dissemination of autonomous vehicles. Finally, our gesture set opens up new research avenues to explore interactive S3D user interfaces in cars. With increasing capabilities of automation, the driver can reclaim time and execute non-driving related tasks. This set offers additional possibilities for the enhanced interaction in cars, while providing a relatively familiar vehicle interior. In summary, this thesis explored various application domains of S3D dashboards in vehicles and answered important questions on viability and performance.

11.1   Summary of Results

- A conservative binocular disparity for content on an S3D dashboard that has a similar layout than ours is D[−0.8° ; +0.58°].
- For an abstract secondary task and with a simple layout, S3D does not necessarily impair, nor improve peripheral change detection or navigation in a list-based menu while driving. Benefits of S3D seem to manifest themselves only in combination with other factors (e.g. with a stimulus that conveys spatial information).
- A virtual agent displayed in S3D does not necessarily improve behavioral or self-reported trust (SAE level 3). However, S3D can increase anthropomorphism of the virtual humanoid agent and anthropomorphism correlates with self-reported trust.

- Performance during time-critical and system-initiated take-over procedures (SAE level 3) can be improved by means of S3D visualizations that show the surroundings. Workload, gaze behavior, and motor reaction time seem to not suffer when applying such visualizations.
- Uni-manual gestures with the right hand are well-suited for the interaction with S3D dashboards. Especially mid-air pointing, mid-air translation and rotation by grabbing virtual or proxy items, and pinch-like scaling showed high agreement rates.

11.2   Recommendations for Future Work

Besides exploring driver distraction, trust, and take-over visualizations further, several other research avenues presented itself over the course of this thesis where S3D displays and dashboard could potentially offer benefits.

11.2.1 Novel and More Accessible Display Types

Results of our studies but also previous research are limited by the applied display technology. Either they were autostereoscopic and provided low visual fidelity or offered high visual fidelity but required stereo glasses. Novel, glasses-free displays like recent light-field (or holographic) displays, could enhance interactive S3D applications in vehicles by offering an experience where the field of view is not restricted by the glasses' frame.

11.2.2 Expand The Scope: Novel Use-Cases

This thesis picked three exemplary challenges in the automotive domain. However, many more exist. Especially the interaction with other road users — motorized or not — is a huge challenge for (semi-)automated vehicles. Here, communicating information about the surroundings with binocular depth cues seems promising since S3D regularly performed good in tasks related to navigation and spatial maps. Further, inspired by our studies, other possible application domains have presented themselves. For example, the usage of such displays for touch-free interaction in automated cabs and public transportation. Such displays could benefit from touch-free, real-direct or proxy interaction to increase hygiene and public health by reducing the transmission of bacteria and viruses. Also, personalization, customization, and situation-aware adaptation of dashboard layouts in general, but also of S3D dashboard with the increased design space, could enhance user experience in vehicles.

11.2.3 Communication

Using S3D, we followed a previous research path to increase anthropomorphism, and by that, trust. We could show that S3D can increase perceived anthropomorphism of visual agents in cars. Also, we could show trust scores and anthropomorphism scores are positively correlated. This correlation paired with the benefits of S3D motivates further research in this area. It could especially be applied and investigated in the domain of human-to-human and human-to-agent collaboration where the communication partner is displayed in S3D. By that, it promises a more anthropomorphic representation and potentially a more trustful information exchange. This, in return, could benefit work-related activities like teleconferencing.

11.2.4 Real-world Driving Studies

The studies of this thesis and the majority of prior work was executed in static low- or medium fidelity driving simulators. However, motion cues are essential for driving — especially when driving performance is a crucial measure. Hence, real-world studies, or at least studies in a high-fidelity driving simulator equipped with a motion platform are necessary to validate results but also to further explore automotive S3D UIs. Here, take-over performance, but also visually induced motion sickness are interesting aspects that research needs to pay attention to.

11.3   Supporting Research Articles

The following articles have been published during the course of this research and support the chapters. For research where I am listed as first author, ideation, conceptualization, implementation, and evaluation were predominantly executed by me. Where I am listed as second author, I supervised students during their thesis. In this case, I supported ideation, conceptualization, implementation, and evaluation.

Christine Haupt, Florian Weidner, and Wolfgang Broll. 2018. personalDash: First Steps Towards User-controlled Personalization of 3D Dashboards with Mobile Devices. In *Proceedings of the 10th International Conference on Automotive User Interfaces and Interactive Vehicular Applications - AutomotiveUI '18*, ACM Press (Ed.). ACM Press, New York, New York, USA, 115–120. https://doi.org/10.1145/3239092.3265952

Kathrin Knutzen, Florian Weidner, and Wolfgang Broll. 2019. Talk to me! Exploring Stereoscopic 3D Anthropomorphic Virtual Assistants in Automated Vehicles. In *Proceedings of the 11th International Conference on Automotive User Interfaces and Interactive Vehicular Applications Adjunct Proceedings - AutomotiveUI '19*. ACM Press, New York, New York, USA, 363–368. https://doi.org/10.1145/3349263.3351503

Florian Weidner and Wolfgang Broll. 2017a. A Spatial Augmented Reality Driving Simulator for Prototyping 3D in-car User Interfaces. In *Proceedings of Driving Simulation Conference 2017 EuropeVR.*

Florian Weidner and Wolfgang Broll. 2017b. Establishing Design Parameters for Large Stereoscopic 3D Dashboards. In *Proceedings of the 9th International Conference on Automotive User Interfaces and Interactive Vehicular Applications Adjunct - AutomotiveUI '17*. ACM Press, New York, New York, USA, 212–216. https://doi.org/10.1145/3131726.3131763

Florian Weidner and Wolfgang Broll. 2019a. Exploring Large Stereoscopic 3D Dashboards for Future Automotive User Interfaces. In *Proceedings of the 9th International Conference on Applied Human Factors and Ergonomics AHFE*. Springer, Cham, Orlando, FL, 502–513. https://doi.org/10.1007/978-3-319-93885-1_45

Florian Weidner and Wolfgang Broll. 2019b. Interact with Your Car: A User-Elicited Gesture Set to Inform Future in-Car User Interfaces. In *Proceedings of the 18th International Conference on Mobile and Ubiquitous Multimedia* (Pisa, Italy) (MUM '19). Association for Computing Machinery, New York, NY, USA, Article 11, 12 pages. https://doi.org/10.1145/3365610.3365625

Florian Weidner and Wolfgang Broll. 2019c. Smart S3D TOR. In *Proceedings of the 8th ACM International Symposium on Pervasive Displays - PerDis '19*. ACM Press, New York, New York, USA, 1–7. https://doi.org/10.1145/3321335.3324937

Florian Weidner and Wolfgang Broll. 2020. Stereoscopic 3D dashboards: An investigation of performance, workload, and gaze behavior during take-overs in semi-autonomous driving. *Personal and Ubiquitous Computing* (aug 2020). https://doi.org/10.1007/s00779-020-01438-8

Florian Weidner, Bernhard Fiedler, Johannes Redlich, Wolfgang Broll, and International Conference on Spatial Audio (ICSA); 5 (Ilmenau): 2019.09.26–28. 2019. Exploring Audiovisual Support Systems for In-Car Multiparty Conferencing. *Audio for Virtual, Augmented and Mixed Realities: Proceedings of ICSA 2019; 5th International Conference on Spatial Audio; September 26th to 28th, 2019, Ilmenau, Germany* (nov 2019), 85–92. https://doi.org/10.22032/dbt.39936

FlorianWeidner, Anne Hoesch, Sandra Poeschl, and Wolfgang Broll. 2017. Comparing VR and non-VR driving simulations: An experimental user study. In *2017 IEEE Virtual Reality (VR)*. IEEE, 281–282. https://doi.org/10.1109/VR.2017. 7892286

Bibliography

Genya Abe, Kenji Sato, and Makoto Itoh. 2018. Driver Trust in Automated Driving Systems: The Case of Overtaking and Passing. *IEEE Transactions on Human-Machine Systems* 48, 1 (feb 2018), 85–94. https://doi.org/10.1109/THMS.2017.2781619

Güliz Acker, Nicolas Schlinkmann, Sophie Käthe Piper, Julia Onken, Peter Vajkoczy, and Thomas Picht. 2018. Stereoscopic Versus Monoscopic Viewing of Aneurysms: Experience of a Single Institution with a Novel Stereoscopic Viewing System. *World Neurosurgery* 119 (2018), 491–501. https://doi.org/10.1016/j.wneu.2018.07.189

Pankaj Aggarwal and Ann L. McGill. 2007. Is That Car Smiling at Me? Schema Congruity as a Basis for Evaluating Anthropomorphized Products. *Journal of Consumer Research* 34, 4 (dec 2007), 468–479. https://doi.org/10.1086/518544

Bashar I. Ahmad, Patrick M. Langdon, Simon J. Godsill, Robert Hardy, Eduardo Dias, and Lee Skrypchuk. 2014. Interactive Displays in Vehicles: Improving Usability with a Pointing Gesture Tracker and Bayesian Intent Predictors. In *Proceedings of the 6th International Conference on Automotive User Interfaces and Interactive Vehicular Applications – AutomotiveUI '14.* 1–8. https://doi.org/10.1145/2667317.2667413

Bashar I. Ahmad, Patrick M. Langdon, Robert Hardy, and Simon J. Godsill. 2015. Intelligent Intent-Aware Touchscreen Systems Using Gesture Tracking with Endpoint Prediction. In *Universal Access in Human-Computer Interaction. Access to Interaction – 9th International Conference, UAHCI 2015, Held as Part of HCI International 2015, Los Angeles, CA, USA, August 2–7, 2015, Proceedings, Part II.* 3–14. https://doi.org/10.1007/978-3-319-20681-3_1

Dohyun Ahn, Youngnam Seo, Minkyung Kim, Joung Huem Kwon, Younbo Jung, Jungsun Ahn, and Doohwang Lee. 2014. The Effects of Actual Human Size Display and Stereoscopic Presentation on Users' Sense of Being Together with and of Psychological Immersion in a Virtual Character. *Cyberpsychology, Behavior, and Social Networking* 17, 7 (jul 2014), 483–487. https://doi.org/10.1089/cyber.2013.0455

Eugene Aidman, Carolyn Chadunow, Kayla Johnson, and John Reece. 2015. Real-time driver drowsiness feedback improves driver alertness and self-reported driving performance. *Accident Analysis & Prevention* 81 (aug 2015), 8–13. https://doi.org/10.1016/j.aap.2015.03.041

Basak Alper, Tobias Hollerer, JoAnn Kuchera-Morin, and Angus Forbes. 2011. Stereoscopic Highlighting: 2D Graph Visualization on Stereo Displays. *IEEE Transactions on Visualization and Computer Graphics* 17, 12 (dec 2011), 2325–2333. https://doi.org/10.1109/TVCG.2011.234

Ignacio Alvarez, Laura Rumbel, and Robert Adams. 2015. Skyline: a Rapid Prototyping Driving Simulator for User Experience. In *Proceedings of the 7th International Conference on Automotive User Interfaces and Interactive Vehicular Applications – AutomotiveUI '15*. ACM Press, New York, New York, USA, 101–108. https://doi.org/10.1145/2799250. 2799290

Roland Arsenault and Colin Ware. 2004. The Importance of Stereo and Eye-Coupled Perspective for Eye-Hand Coordination in Fish Tank VR. *Presence: Teleoperators and Virtual Environments* 13, 5 (oct 2004), 549–559. https://doi.org/10.1162/1054746042545300

Kevin W. Arthur, Kellogg S. Booth, and Colin Ware. 1993. Evaluating 3D task performance for fish tank virtual worlds. *ACM Transactions on Information Systems* 11, 3 (1993), 239–265. https://doi.org/10.1145/159161.155359

Jackie Ayoub, Feng Zhou, Shan Bao, and X. Jessie Yang. 2019. From Manual Driving to Automated Driving: A Review of 10 Years of AutoUI. In *Proceedings of the 11th International Conference on Automotive User Interfaces and Interactive Vehicular Applications – AutomotiveUI '19*. ACM Press, New York, New York, USA, 70–90. https://doi.org/10. 1145/3342197.3344529

Pavlo Bazilinskyy, Sebastiaan Petermeijer, Veronika Petrovych, Dimitra Dodou, and Joost de Winter. 2018. Take-over requests in highly automated driving: A crowdsourcing survey on auditory, vibrotactile, and visual displays. *Transportation Research Part F: Traffic Psychology and Behaviour* 56, February (jul 2018), 82–98. https://doi.org/10.1016/j.trf.2018. 04.001

Simone Benedetto, Marco Pedrotti, Luca Minin, Thierry Baccino, Alessandra Re, and Roberto Montanari. 2011. Driver workload and eye blink duration. *Transportation Research Part F: Traffic Psychology and Behaviour* 14, 3 (2011), 199–208. https://doi.org/10.1016/j.trf. 2010.12.001

Frank Beruscha, Wolfgang Krautter, Anja Lahmer, and Markus Pauly. 2017. An evaluation of the influence of haptic feedback on gaze behavior during in-car interaction with touch screens. In *2017 IEEE World Haptics Conference (WHC)*. IEEE, 201–206. https://doi.org/ 10.1109/WHC.2017.7989901

Oliver Bimber and Ramesh Raskar. 2005. *Spatial Augmented Reality Merging Real and Virtual Worlds* (1st ed.). Vol. 6. A K Peters, Ltd., Wellesley, MA. https://doi.org/10.1260/ 147807708784640126

Susanne Boll, Andrew L. Kun, Andreas Riener, and C. Y. David Yang. 2019. Users and automated driving systems: How will we interact with tomorrow's vehicles? (Report from Dagstuhl Seminar 19132). *Dagstuhl Reports* 9, 3 (2019), 111–178. https://doi.org/10.4230/ DagRep.9.3.111

Shadan Sadeghian Borojeni, Susanne Boll, Wilko Heuten, Heinrich Bülthoff, and Lewis Chuang. 2018. Feel the Movement: Real Motion Influences Responses to Take-over Requests in Highly Automated Vehicles. *Proceedings of the 2018 CHI Conference on Human Factors in Computing Systems – CHI '18* (2018), 1–13. https://doi.org/10.1145/ 3173574.3173820

Shadan Sadeghian Borojeni, Lewis Chuang, Wilko Heuten, and Susanne Boll. 2016. Assisting Drivers with Ambient Take-Over Requests in Highly Automated Driving. In *Proceedings of the 8th International Conference on Automotive User Interfaces and Interactive Vehicular Applications – Automotive'UI 16*. ACM Press, New York, New York, USA, 237–244. https://doi.org/10.1145/3003715.3005409

Shadan Sadeghian Borojeni, Torben Wallbaum, Wilko Heuten, and Susanne Boll. 2017. Comparing Shape-Changing and Vibro-Tactile Steering Wheels for Take-Over Requests in Highly Automated Driving. In *Proceedings of the 9th International Conference on Automotive User Interfaces and Interactive Vehicular Applications – AutomotiveUI '17*. ACM Press, New York, New York, USA, 221–225. https://doi.org/10.1145/3122986.3123003

Sabah Boustila, Dominique Bechmann, and Antonio Capobianco. 2017. Effects of stereo and head tracking on distance estimation, presence, and simulator sickness using wall screen in architectural project review. In *2017 IEEE Symposium on 3D User Interfaces (3DUI)*. IEEE, 231–232. https://doi.org/10.1109/3DUI.2017.7893356

Elizabeth Broadbent, Vinayak Kumar, Xingyan Li, John Sollers, Rebecca Q. Stafford, Bruce A. MacDonald, and Daniel M. Wegner. 2013. Robots with Display Screens: A Robot with a More Humanlike Face Display Is Perceived To Have More Mind and a Better Personality. *PLOS ONE* 8, 8 (aug 2013). https://doi.org/10.1371/journal.pone.0072589

Morton B. Brown and Alan B. Forsythe. 1974. Robust Tests for the Equality of Variances. *J. Amer. Statist. Assoc.* 69, 346 (jun 1974), 364–367. https://doi.org/10.1080/01621459.1974.10482955

Nora Broy. 2016. *Stereoscopic 3D User Interfaces: Exploring the Potentials and Risks of 3D Displays in Cars*. Dissertation. Universität Stuttgart. https://doi.org/10.18419/opus-8851

Nora Broy, Florian Alt, Stefan Schneegass, Niels Henze, and Albrecht Schmidt. 2013. Perceiving layered information on 3D displays using binocular disparity. *Proceedings of the 2nd ACM International Symposium on Pervasive Displays – PerDis '13* (2013), 61. https://doi.org/10.1145/2491568.2491582

Nora Broy, Florian Alt, Stefan Schneegass, and Bastian Pfleging. 2014a. 3D Displays in Cars: Exploring the User Performance for a Stereoscopic Instrument Cluster. In *Proceedings of the 6th International Conference on Automotive User Interfaces and Interactive Vehicular Applications – AutomotiveUI '14*. 1–9. https://doi.org/10.1145/2667317.2667319

Nora Broy, Elisabeth André, and Albrecht Schmidt. 2012. Is stereoscopic 3D a better choice for information representation in the car?. In *Proceedings of the 4th International Conference on Automotive User Interfaces and Interactive Vehicular Applications – AutomotiveUI '12*. https://doi.org/10.1145/2390256.2390270

Nora Broy, Mengbing Guo, Stefan Schneegass, Bastian Pfleging, and Florian Alt. 2015a. Introducing novel technologies in the car – Conducting a Real-World Study to Test 3D Dashboards. In *Proceedings of the 7th International Conference on Automotive User Interfaces and Interactive Vehicular Applications – AutomotiveUI '15*. ACM Press, New York, New York, USA, 179–186. https://doi.org/10.1145/2799250.2799280

Nora Broy, Stefan Schneegass, Mengbing Guo, Florian Alt, and Albrecht Schmidt. 2015b. Evaluating Stereoscopic 3D for Automotive User Interfaces in a Real-World Driving Study. In *Extended Abstracts of the ACM CHI'15 Conference on Human Factors in Computing Systems*. https://doi.org/10.1145/2702613.2732902

Nora Broy, Benedikt Zierer, Stefan Schneegass, and Florian Alt. 2014b. Exploring virtual depth for automotive instrument cluster concepts. In *Proceedings of the extended abstracts of the 32nd annual ACM conference on Human factors in computing systems – CHI EA '14*. 1783–1788. https://doi.org/10.1145/2559206.2581362

Gerd Bruder, Frank Steinicke, and Wolfgang Sturzlinger. 2013. To touch or not to touch?. In *Proceedings of the 1st symposium on Spatial user interaction – SUI '13 (SUI '13)*. ACM Press, New York, New York, USA, 9. https://doi.org/10.1145/2491367.2491369

Katia Buchhop, Laura Edel, Sabrin Kenaan, Ulrike Raab, Patricia Böhm, and Daniel Isemann. 2017. In-Vehicle Touchscreen Interaction: Can a Head-Down Display Give a Heads-Up on Obstacles on the Road? *Proceedings of the 9th International Conference on Automotive User Interfaces and Interactive Vehicular Applications – AutomotiveUI '17* (2017), 21–30. https://doi.org/10.1145/3122986.3123001

M. Bueno, Ebru Dogan, F. Hadj Selem, E. Monacelli, S. Boverie, and A. Guillaume. 2016. How different mental workload levels affect the take-over control after automated driving. In *2016 IEEE 19th International Conference on Intelligent Transportation Systems (ITSC)*. IEEE, 2040–2045. https://doi.org/10.1109/ITSC.2016.7795886

Timothy J. Buker, Dennis A. Vincenzi, and John E. Deaton. 2012. The effect of apparent latency on simulator sickness while using a see-through helmet-mounted display: Reducing apparent latency with predictive compensation. *Human Factors* 54, 2 (2012), 235–249. https://doi.org/10.1177/0018720811428734

Bundesanstalt für Straßenwesen. 2017. Aktualisierung des Überholmodells auf Landstraßen. https://www.bast.de/BASt_2017/DE/Publikationen/Foko/2017-2016/2017-05.html

Totsapon Butmee, Terry C. Lansdown, and Guy H. Walker. 2019. *Mental Workload and Performance Measurements in Driving Task: A Review Literature*. Vol. 818. Springer International Publishing. 286–294 pages. https://doi.org/10.1007/978-3-319-96098-2

Jeff K. Caird and William J. Horrey. 2011. Twelve Practical And Useful Questions About Driving Simulation. In *Handbook of Driving Simulation for Engineering, Medicine, and Psychology* (1st ed.), Donald L. Fisher, Matthew Rizzo, Jeff K. Caird, and John D. Lee (Eds.). CRC Press, Boca Raton, FL, Chapter 5, 77–94.

Erion Çano, Riccardo Coppola, Eleonora Gargiulo, Marco Marengo, and Maurizio Morisio. 2017. Mood-Based On-Car Music Recommendations. In *Industrial Networks and Intelligent Systems. INISCOM 2016. Lecture Notes of the Institute for Computer Sciences, Social Informatics and Telecommunications Engineering*, L. Maglaras, H. Janicke, and K. Jones (Eds.), Vol. 188. 154–163. https://doi.org/10.1007/978-3-319-52569-3_14

Rikk Carey and Gavin Bell. 1997. *The Annotated VRML 2.0 Reference Manual. Addison-Wesley Longman Ltd.*, GBR. http://www.x-3-x.net/vrml/archive/annotatedVRML2/BOOK.HTM

Rebecca L. Charles and Jim Nixon. 2019. Measuring mental workload using physiological measures: A systematic review. *Applied Ergonomics* 74, January 2019 (2019), 221–232. https://doi.org/10.1016/j.apergo.2018.08.028

Enguo Chen, Jing Cai, Xiangyao Zeng, Sheng Xu, Yun Ye, Qun Frank Yan, and Tailiang Guo. 2019. Ultra-large moiré-less autostereoscopic three-dimensional light-emitting-diode displays. *Optics Express* 27, 7 (apr 2019), 10355. https://doi.org/10.1364/OE.27.010355

Jessie. Y. C. Chen, Razia. N. V. Oden, Caitlin Kenny, and John O. Merritt. 2010. Stereoscopic Displays for Robot Teleoperation and Simulated Driving. *Proceedings of the Human Factors and Ergonomics Society Annual Meeting* 54, 19 (2010), 1488–1492. https://doi.org/10.1177/154193121005401928

Alexey Chistyakov and Jordi Carrabina. 2015. An HTML Tool for Production of Interactive Stereoscopic Content. In *Ubiquitous Computing and Ambient Intelligence. Sensing, Processing, and Using Environmental Information*. Vol. 9454. Springer International Publishing Switzerland, 449–459. https://doi.org/10.1007/978-3-319-26401-1_42

Song-Woo Choi, Siyeong Lee, Min-Woo Seo, and Suk-Ju Kang. 2018. Time Sequential Motion-to-Photon Latency Measurement System for Virtual Reality Head-Mounted Displays. *Electronics* 7, 9 (2018), 171. https://doi.org/10.3390/electronics7090171

Joon Hao Chuah, Andrew Robb, Casey White, Adam Wendling, Samsun Lampotang, Regis Kopper, and Benjamin Lok. 2013. Exploring Agent Physicality and Social Presence for Medical Team Training. *Presence: Teleoperators and Virtual Environments* 22, 2 (aug 2013), 141–170. https://doi.org/10.1162/PRES_a_00145

Jan Conrad, Dieter Wallach, Arthur Barz, Daniel Kerpen, Tobias Puderer, and Andreas Weisenburg. 2019. Concept simulator K3F: A Flexible Framework for Driving Simulations. In *Proceedings of the 11th International Conference on Automotive User Interfaces and Interactive Vehicular Applications Adjunct Proceedings – AutomotiveUI '19*. ACM Press, New York, New York, USA, 498–501. https://doi.org/10.1145/3349263.3349596

Continental AG. 2019. Continental and Leia's new 3D Lightfield Display bring the third dimension into Automotive vehicles. https://www.continental.com/en/press/press-releases/2019-06-11-3d-instrument-cluster-174836

Kimberly E. Culley and Poornima Madhavan. 2013. A note of caution regarding anthropomorphism in HCI agents. *Computers in Human Behavior* 29, 3 (may 2013), 577–579. https://doi.org/10.1016/j.chb.2012.11.023

James E Cutting. 1997. How the eye measures reality and virtual reality. *Behavior Research Methods, Instruments, & Computers* 29, 1 (mar 1997), 27–36. https://doi.org/10.3758/BF03200563

Rita Cyganski, Eva Fraedrich, and Barbara Lenz. 2015. Travel-time valuation for automated driving: A use-case-driven study. In *Proceedings of the 94th Annual Meeting of the TRB*. https://elib.dlr.de/95260/

Leonard Daly and Don Brutzman. 2007. X3D: Extensible 3D Graphics Standard [Standards in a Nutshell]. *IEEE Signal Processing Magazine* 24, 6 (nov 2007), 130–135. https://doi.org/10.1109/MSP.2007.905889

Celso M. de Melo, Peter J. Carnevale, Stephen J. Read, and Jonathan Gratch. 2014. Reading people's minds from emotion expressions in interdependent decision making. *Journal of Personality and Social Psychology* 106, 1 (2014), 73–88. https://doi.org/10.1037/a0034251

Lucio Tommaso De Paolis and Antonio Mongelli. 2014. Stereoscopic-3D Vision to Improve Situational Awareness in Military Operations. *Lecture Notes in Computer Science (including subseries Lecture Notes in Artificial Intelligence and Lecture Notes in Bioinformatics)* 8853 (2014), 351–362. https://doi.org/10.1007/978-3-319-13969-2

Ewart J. de Visser, Frank Krueger, Patrick McKnight, Steven Scheid, Melissa Smith, Stephanie Chalk, and Raja Parasuraman. 2012. The World is not Enough: Trust in Cognitive Agents. *Proceedings of the Human Factors and Ergonomics Society Annual Meeting* 56, 1 (sep 2012), 263–267. https://doi.org/10.1177/1071181312561062

Ewart J. de Visser, Samuel S. Monfort, Ryan McKendrick, Melissa A. B. Smith, Patrick E. McKnight, Frank Krueger, and Raja Parasuraman. 2016. Almost human: Anthropomorphism increases trust resilience in cognitive agents. *Journal of Experimental Psychology: Applied* 22, 3 (sep 2016), 331–349. https://doi.org/10.1037/xap0000092

Matthieu Deru and Robert Neßelrath. 2015. autoUI-ML: A design language for the flexible creation of automotive GUIs based on semantically represented data. In *Proceedings of the 4th International Symposium on Pervasive Displays – PerDis '15*. ACM Press, New York, New York, USA, 235–236. https://doi.org/10.1145/2757710.2776799

Andre Dettmann and Angelika C. Bullinger. 2019. Autostereoscopic Displays for In-Vehicle
Applications. In *Proceedings of the 20th Congress of the International Ergonomics Association (IEA 2018) (Advances in Intelligent Systems and Computing, Vol. 821)*, Sebas-
tiano Bagnara, Riccardo Tartaglia, Sara Albolino, Thomas Alexander, and Yushi Fujita
(Eds.). Springer International Publishing, Cham, 457–466. https://doi.org/10.1007/978-3-
319-96080-7

Andre Dettmann and Angelika C. Bullinger. 2020. Investigation on the Effectiveness of
Autostereoscopic 3D Displays for Parking Maneuver Tasks with Passenger Cars. *Advances
in Intelligent Systems and Computing* 964 (2020), 608–617. https://doi.org/10.1007/978-
3-030-20503-4_55

Paul E. Dickson, Jeremy E. Block, Gina N. Echevarria, and Kristina C. Keenan. 2017. An
experience-based comparison of unity and unreal for a stand-alone 3D game development
course. *Annual Conference on Innovation and Technology in Computer Science Education,
ITiCSE* Part F1286 (2017), 70–75. https://doi.org/10.1145/3059009.3059013

Sharon Dixon, Elisabeth Fitzhugh, and Denise Aleva. 2009. Human factors guidelines for
applications of 3D perspectives: a literature review. In *Proc. SPIE 7327, Display Technologies and Applications for Defense, Security, and Avionics III*, John T. Thomas and Daniel D.
Desjardins (Eds.). https://doi.org/10.1117/12.820853

Andrew T. Duchowski, Krzysztof Krejtz, Izabela Krejtz, Cezary Biele, Anna Niedzielska,
Peter Kiefer, Martin Raubal, and Ioannis Giannopoulos. 2018. The Index of Pupillary
Activity. In *Proceedings of the 2018 CHI Conference on Human Factors in Computing
Systems – CHI '18*. ACM Press, New York, New York, USA, 1–13. https://doi.org/10.
1145/3173574.3173856

Joshua E. Ekandem, Ignacio Alvarez, Cat Rayburn, and Andrea Johnson. 2018. Theo, take a
right ... uh ... left: Conversational Route Negotiations with Autonomous Driving Assistants.
In *Proceedings of the 10th International Conference on Automotive User Interfaces and
Interactive Vehicular Applications – AutomotiveUI '18*. ACM Press, New York, New York,
USA, 261–262. https://doi.org/10.1145/3239092.3267414

Alan Elliott and Wayne Woodward. 2007. *Statistical Analysis Quick Reference Guidebook*.
SAGE Publications, Inc., 2455 Teller Road, Thousand Oaks California 91320 United States
of America. https://doi.org/10.4135/9781412985949

David G Elmes, Barry H Kantowitz, and Henry L Roediger. 2012. *Research methods in
psychology*. Wadsworth Cengage Learning, [Melbourne, Victoria].

Mica R. Endsley. 1995. Toward a Theory of Situation Awareness in Dynamic Systems. *Human
Factors: The Journal of the Human Factors and Ergonomics Society* 37, 1 (1995), 32–64.
https://doi.org/10.1518/001872095779049543

Mica R. Endsley. 2017. From Here to Autonomy: Lessons Learned from Human-Automation
Research. *Human Factors* 59, 1 (2017), 5–27. https://doi.org/10.1177/0018720816681350

Mica R. Endsley and Esin O. Kiris. 1995. The Out-of-the-Loop Performance Problem
and Level of Control in Automation. *Human Factors: The Journal of the Human
Factors and Ergonomics Society* 37, 2 (jun 1995), 381–394. https://doi.org/10.1518/
001872095779064555

Mario Enriquez, Oleg Afonin, Brent Yager, and Karon Maclean. 2001. A Pneumatic Tactile Alerting System for the Driving Environment. *Proceedings of the 2001 workshop on
Perceptive user interfaces – PUI '01* (2001), 1–7. https://doi.org/10.1145/971478.971506

Friederike Eyssel and Steve Loughnan. 2013. "It Don't Matter If You're Black or White"?. In *International Conference on Social Robotics*, Guido Herrmann, Martin J. Pearson, Alexander Lenz, Paul Bremner, Adam Spiers, and Ute Leonards (Eds.). Springer, Cham, Bristol, 422–431. https://doi.org/10.1007/978-3-319-02675-6_42

Hessam Jahani Fariman, Hasan J. Alyamani, Manolya Kavakli, and Len Hamey. 2016. Designing a user-defined gesture vocabulary for an in-vehicle climate control system. In *Proceedings of the 28th Australian Conference on Computer-Human Interaction – OzCHI '16*. ACM Press, New York, New York, USA, 391–395. https://doi.org/10.1145/3010915.3010955

Pablo Figueroa, Omer Medina, Roger Jiménez, José Martinez, and Camilo Albarracin. 2005. Extensions for Interactivity and Retargeting in X3D. In *Proceedings of the Tenth International Conference on 3D Web Technology (Web3D '05)*. Association for Computing Machinery, New York, NY, USA, 103–110. https://doi.org/10.1145/1050491.1050506

Donald L. Fisher and Matthew Rizzo. 2011. *Handbook of Driving Simulation for Engineering, Medicine and Psychology*. CRC Press, Boca Raton. 273–294 pages.

Yannick Forster, Svenja Paradies, and Nikolaus Bee. 2015. The third dimension: Stereoscopic displaying in a fully immersive driving simulator. In *Proceedings of DSC 2015 Europe Driving Simulation Conference & Exhibition*, Heinrich Bülthoff, Andras Kemeny, and Paolo Pretto (Eds.). 25–32.

Anna-katharina Frison, Philipp Wintersberger, Tianjia Liu, and Andreas Riener. 2019. Why do you like to drive automated?. In *Proceedings of the 24th International Conference on Intelligent User Interfaces – IUI '19*. ACM Press, New York, New York, USA, 528–537. https://doi.org/10.1145/3301275.3302331

Joseph L. Gabbard, Missie Smith, Kyle Tanous, Hyungil Kim, and Bryan Jonas. 2019. AR DriveSim: An Immersive Driving Simulator for Augmented Reality Head-Up Display Research. *Frontiers in Robotics and AI* 6, October 2019 (oct 2019), 1–16. https://doi.org/10.3389/frobt.2019.00098

Andrea Gaggioli and Ralf Breining. 2001. Perception and cognition in immersive Virtual Reality. In *Communications Through Virtual Technology: Identity Community and Technology in the Internet Age*. IOS Press, 71–86.

Mark A. Georgeson and Stuart A. Wallis. 2014. Binocular fusion, suppression and diplopia for blurred edges. *Ophthalmic & physiological optics: the journal of the British College of Ophthalmic Opticians (Optometrists)* 34, 2 (mar 2014), 163–185. https://doi.org/10.1111/opo.12108

Michael A Gerber, Ronald Schroeter, and Julia Vehns. 2019. A Video-Based Automated Driving Simulator for Automotive UI Prototyping, UX and Behaviour Research. In *Proceedings of the 11th International Conference on Automotive User Interfaces and Interactive Vehicular Applications – AutomotiveUI '19*. ACM Press, New York, New York, USA, 14–23. https://doi.org/10.1145/3342197.3344533

Christian Gold, Ilirjan Berisha, and Klaus Bengler. 2015. Utilization of drivetime – Performing non-driving related tasks while driving highly automated. In *Proceedings of the Human Factors and Ergonomics Society*. https://doi.org/10.1177/1541931215591360

Christian Gold, Daniel Damböck, Lutz Lorenz, and Klaus Bengler. 2013. "Take over!" How long does it take to get the driver back into the loop? *Proceedings of the Human Factors and Ergonomics Society Annual Meeting* 57, 1 (2013), 1938–1942. https://doi.org/10.1177/1541931213571433

Bruce E. Goldstein. 2014. *Sensation and Perception* (9th ed.). Wadsworth, Belmont, CA, USA. 489 pages. www.cengagebrain.com

Diego Gonzalez-Zuniga and Jordi Carrabina. 2015. S3doodle: Case Study for Stereoscopic Gui and user Content Creation. *International Conference on Communication, Media, Technology and Design (ICCMTD)* (2015), 129–137. http://www.cmdconf.net/2015/pdf/36.pdf

Diego Gonzalez-Zuniga, Toni Granollers, and Jordi Carrabina. 2015. Sketching Stereoscopic GUIs with HTML5 Canvas. In *UCAmI 2015: Ubiquitous Computing and Ambient Intelligence. Sensing, Processing, and Using Environmental Information*, Vol. 9454. 289–296. https://doi.org/10.1007/978-3-319-26401-1_27

Daniel Gotsch, Xujing Zhang, Timothy Merritt, and Roel Vertegaal. 2018. TeleHuman2. In *Proceedings of the 2018 CHI Conference on Human Factors in Computing Systems – CHI '18*. ACM Press, New York, New York, USA, 1–10. https://doi.org/10.1145/3173574.3174096

Nikhil Gowda, Kirstin Kohler, and Wendy Ju. 2014. Dashboard Design for an Autonomous Car. *Proceedings of the 6th International Conference on Automotive User Interfaces and Interactive Vehicular Applications – AutomotiveUI '14* (2014), 1–4. https://doi.org/10.1145/2667239.2667313

Jeffrey Greenberg and Mike Blommer. 2011. Physical Fidelity of Driving Simulators. In *Handbook of Driving Simulation for Engineering, Medicine, and Psychology* (1st ed.), Donald L. Fisher, Matthew Rizzo, Jeff K. Caird, and John D. Lee (Eds.). CRC Press, Boca Raton, FL, 108–131.

Celeste Groenewald, Craig Anslow, Junayed Islam, Chris Rooney, Peter Passmore, and William Wong. 2016. Understanding 3D Mid-Air Hand Gestures with Interactive Surfaces and Displays: A Systematic Literature Review. In *HCI '16 Proceedings of the 30th International BCS Human Computer Interaction Conference: Fusion!* 1–13. https://doi.org/10.14236/ewic/HCI2016.43

Martin Hachet, Benoit Bossavit, Aurélie Cohé, and Jean-Baptiste de la Rivière. 2011. Toucheo: Multitouch and Stereo Combined in a Seamless Workspace. In *Proceedings of the 24th Annual ACM Symposium on User Interface Software and Technology (UIST '11)*. ACM, New York, NY, USA, 587–592. https://doi.org/10.1145/2047196.2047273

Jonna Häkkilä, Ashley Colley, and Juho Rantakari. 2014. Exploring Mixed Reality Window Concept for Car Passengers. In *Proceedings of the 6th International Conference on Automotive User Interfaces and Interactive Vehicular Applications – AutomotiveUI '14*. 1–4. https://doi.org/10.1145/2667239.2667288

Peter A. Hancock, Deborah R. Billings, Kristin E. Schaefer, Jessie Y.C. Chen, Ewart J. De Visser, and Raja Parasuraman. 2011. A meta-analysis of factors affecting trust in human-robot interaction. *Human Factors* 53, 5 (2011), 517–527. https://doi.org/10.1177/0018720811417254

Sandra G. Hart and Lowell E. Staveland. 1988. Development of NASA-TLX (Task Load Index): Results of Empirical and Theoretical Research. In *Human Mental Workload*, Peter A Hancock and Najmedin Meshkati (Eds.). Advances in Psychology, Vol. 52. North-Holland, 139–183. https://doi.org/10.1016/S0166-4115(08)62386-9

Joerg Hauber. 2012. The Impact of Collaborative Style on the Perception of 2D and 3D Videoconferencing Interfaces. *The Open Software Engineering Journal* 6, 1 (jun 2012), 1–20. https://doi.org/10.2174/1874107X01206010001

Jörg Hauber, Holger Regenbrecht, Mark Billinghurst, and Andy Cockburn. 2006. Spatiality in videoconferencing. In *Proceedings of the 2006 20th anniversary conference on Computer supported cooperative work – CSCW '06*. ACM Press, New York, New York, USA, 413. https://doi.org/10.1145/1180875.1180937

Sabrina M. Hegner, Ardion D. Beldad, and Gary J. Brunswick. 2019. In Automatic We Trust: Investigating the Impact of Trust, Control, Personality Characteristics, and Extrinsic and Intrinsic Motivations on the Acceptance of Autonomous Vehicles. *International Journal of Human-Computer Interaction* 35, 19 (2019), 1769–1780. https://doi.org/10.1080/10447318.2019.1572353

Sheila S. Hemami, Francis M. Ciaramello, Sean S. Chen, Nathan G. Drenkow, Dae Yeol Lee, Seonwoo Lee, Evan G. Levine, and Adam J. McCann. 2012. Comparing user experiences in 2D and 3D videoconferencing. *Proceedings – International Conference on Image Processing, ICIP* (2012), 1969–1972. https://doi.org/10.1109/ICIP.2012.6467273

Nicola Herbison, Isabel M. Ash, Daisy MacKeith, Anthony Vivian, Jonathan H. Purdy, Apostolos Fakis, Sue V. Cobb, Trish Hepburn, Richard M. Eastgate, Richard M. Gregson, and Alexander J. E. Foss. 2015. Interactive stereo games to improve vision in children with amblyopia using dichoptic stimulation. In *Stereoscopic Displays and Applications XXVI*, Nicolas S. Holliman, Andrew J. Woods, Gregg E. Favalora, and Takashi Kawai (Eds.), Vol. 9391. 93910A. https://doi.org/10.1117/12.2083360

R. Bruce Herrick. 1968. Rear Projection Screens: a Theoretical Analysis. *Applied Optics* 7, 5 (may 1968), 763. https://doi.org/10.1364/AO.7.000763

Jahani F. Hessam, Massimo Zancanaro, Manolya Kavakli, and Mark Billinghurst. 2017. Towards optimization of mid-air gestures for in-vehicle interactions. In *ACM International Conference Proceeding Series*. 126–134. https://doi.org/10.1145/3152771.3152785

Philipp Hock, Johannes Kraus, Marcel Walch, Nina Lang, and Martin Baumann. 2016. Elaborating Feedback Strategies for Maintaining Automation in Highly Automated Driving. *Proceedings of the 8th International Conference on Automotive User Interfaces and Interactive Vehicular Applications – Automotive'UI 16* (2016), 105–112. https://doi.org/10.1145/3003715.3005414

Anne Hoesch, Sandra Poeschl, Florian Weidner, and Nicola Doering. 2016. Effects of Fidelity on Driving Behavior – An Experimental Study. In *21th Annual CyperPsychology, Cyper-Therapy & Social Networking Conference (CyPsy21)*.

Anne Hoesch, Sandra Poeschl, Florian Weidner, Roberto Walter, and Nicola Doering. 2018. The Relationship between Visual Attention and Simulator Sickness: A Driving Simulation Study. *2018 IEEE Virtual Reality (VR)* (2018). https://doi.org/10.1109/VR

Kevin Anthony Hoff and Masooda Bashir. 2015. Trust in Automation. *Human Factors: The Journal of the Human Factors and Ergonomics Society* 57, 3 (may 2015), 407–434. https://doi.org/10.1177/0018720814547570

Ekram Hossain and Arun K. Kulshreshth. 2019. Exploring the Effects of Stereoscopic 3D on Gaming Experience Using Physiological Sensors. In *Symposium on Spatial User Interaction*. ACM, New York, NY, USA, 1–2. https://doi.org/10.1145/3357251.3358751

Ian P. Howard and Brian J. Rogers. 2008. *Binocular Vision and Stereopsis*. Oxford University Press, New York, New York, USA. 1–746 pages. https://doi.org/10.1093/acprof:oso/9780195084764.001.0001 arxiv:arXiv:1011.1669v3

Wijnand Ijsselsteijn, Huib De Ridder, Jonathan Freeman, S. E. Avons, and Don Bouwhuis. 2001. Effects of stereoscopic presentation, image motion, and screen size on subjective and objective corroborative measures of presence. *Presence: Teleoperators and Virtual Environments* 10, 3 (2001), 298–311. https://doi.org/10.1162/105474601300343621

Ohad Inbar and Noam Tractinsky. 2011. Make a Trip an Experience: Sharing In-Car Information with Passengers. *Chi'11* (2011), 1243–1248. https://doi.org/10.1145/1979742.1979755

Christian P. Janssen, Andrew L. Kun, Stephen Brewster, Linda Ng Boyle, Duncan P. Brumby, and Lewis L. Chuang. 2019. Exploring the concept of the (future) mobile office. In *Proceedings of the 11th International Conference on Automotive User Interfaces and Interactive Vehicular Applications Adjunct Proceedings – AutomotiveUI '19*. ACM Press, New York, New York, USA, 465–467. https://doi.org/10.1145/3349263.3349600

Jason Jerald. 2016. *The VR Book* (1st ed.). Morgan & Claypool Publishers All.

Jiun-Yin Jian, Ann M. Bisantz, and Colin G. Drury. 2000. Foundations for an Empirically Determined Scale of Trust in Automated Systems. *International Journal of Cognitive Ergonomics* 4, 1 (mar 2000), 53–71. https://doi.org/10.1207/S15327566IJCE0401_04

Sofia Jorlöv, Katarina Bohman, and Annika Larsson. 2017. Seating Positions and Activities in Highly Automated Cars – A Qualitative Study of Future Automated Driving Scenarios. *IRCOBI Conference* (2017), 13–22. http://www.ircobi.org/wordpress/downloads/irc17/pdf-files/11.pdf

Deborah Kendzierski and Kenneth J. DeCarlo. 2016. Physical Activity Enjoyment Scale: Two Validation Studies. *Journal of Sport and Exercise Psychology* 13, 1 (2016), 50–64. https://doi.org/10.1123/jsep.13.1.50

Robert S. Kennedy, Norman E. Lane, Kevin S. Berbaum, and Michael G. Lilienthal. 1993. Simulator Sickness Questionnaire: An Enhanced Method for Quantifying Simulator Sickness. *The International Journal of Aviation Psychology* 3, 3 (jul 1993), 203–220. https://doi.org/10.1207/s15327108ijap0303_3

Dagmar Kern and Bastian Pfleging. 2013. Supporting interaction through haptic feedback in automotive user interfaces. *Interactions* 20, 2 (2013), 16. https://doi.org/10.1145/2427076.2427081

Dagmar Kern and Albrecht Schmidt. 2009. Design space for driver-based automotive user interfaces. In *Proceedings of the 1st International Conference on Automotive User Interfaces and Interactive Vehicular Applications – AutomotiveUI '09*. https://doi.org/10.1145/1620509.1620511

Sumbul Khan and Bige Tunçer. 2019. Gesture and speech elicitation for 3D CAD modeling in conceptual design. *Automation in Construction* 106, May (2019), 102847. https://doi.org/10.1016/j.autcon.2019.102847

Konstantina Kilteni, Raphaela Groten, and Mel Slater. 2012. The Sense of Embodiment in Virtual Reality. *Presence: Teleoperators and Virtual Environments* 21, 4 (nov 2012), 373–387. https://doi.org/10.1162/PRES_a_00124

Kibum Kim, John Bolton, Audrey Girouard, Jeremy Cooperstock, and Roel Vertegaal. 2012. TeleHuman. In *Proceedings of the 2012 ACM annual conference on Human Factors in Computing Systems – CHI '12*. ACM Press, New York, New York, USA, 2531. https://doi.org/10.1145/2207676.2208640

Naeun Kim, Kwangmin Jeong, Minyoung Yang, Yejeon Oh, and Jinwoo Kim. 2017. Are You Ready to Take-over?. In *Proceedings of the 2017 CHI Conference Extended Abstracts on Human Factors in Computing Systems – CHI EA '17*. ACM Press, New York, New York, USA, 1771–1778. https://doi.org/10.1145/3027063.3053155

Eugenia M. Kolasinski. 1995. *Simulator Sickness in Virtual Environments*. Technical Report. U.S. Army Research Institute for the Behavioral and Social Sciences, Alexandria, VA. 68 pages. https://apps.dtic.mil/dtic/tr/fulltext/u2/a295861.pdf

Sirisilp Kongsilp and Matthew N. Dailey. 2017. Motion parallax from head movement enhances stereoscopic displays by improving presence and decreasing visual fatigue. *Displays* 49 (2017), 72–79. https://doi.org/10.1016/j.displa.2017.07.001

Jeamin Koo, Jungsuk Kwac, Wendy Ju, Martin Steinert, Larry Leifer, and Clifford Nass. 2015. Why did my car just do that? Explaining semi-autonomous driving actions to improve driver understanding, trust, and performance. *International Journal on Interactive Design and Manufacturing (IJIDeM)* 9, 4 (nov 2015), 269–275. https://doi.org/10.1007/s12008-014-0227-2

Frank L. Kooi, Daan Dekker, Raymond Van Ee, and Anne Marie Brouwer. 2010. Real 3D increases perceived depth over anaglyphs but does not cancel stereo-anomaly. *Displays* 31, 3 (2010), 132–138. https://doi.org/10.1016/j.displa.2010.03.003

Anne Köpsel and Nikola Bubalo. 2015. Benefiting from legacy bias. *Interactions* 22, 5 (2015), 44–47. https://doi.org/10.1145/2803169

Arthur F. Kramer. 1990. *Physiological Metrics of Mental Workload: A Review of Recent Progress*. Technical Report. University of Illinois Urbana-Champaign. https://apps.dtic.mil/docs/citations/ADA223701

Johannes Maria Kraus, Florian Nothdurft, Philipp Hock, David Scholz, Wolfgang Minker, and Martin Baumann. 2016. Human After All. In *Proceedings of the 8th International Conference on Automotive User Interfaces and Interactive Vehicular Applications Adjunct – Automotive'UI 16*. ACM Press, New York, New York, USA, 129–134. https://doi.org/10.1145/3004323.3004338

Krzysztof Krejtz, Andrew T. Duchowski, Anna Niedzielska, Cezary Biele, and Izabela Krejtz. 2018. Eye tracking cognitive load using pupil diameter and microsaccades with fixed gaze. *PLOS ONE* 13, 9 (sep 2018), e0203629. https://doi.org/10.1371/journal.pone.0203629

Karen Krüger. 2008. *Nutzen und Grenzen von 3D-Anzeigen in Fahrzeugen*. Dissertation. Humboldt-Universität zu Berlin. http://edoc.hu-berlin.de/dissertationen/krueger-karen-2007-11-09/PDF/krueger.pdf

Philipp Kulms and Stefan Kopp. 2019. More Human-Likeness, More Trust?. In *Proceedings of Mensch und Computer 2019 on – MuC'19*. ACM Press, New York, New York, USA, 31–42. https://doi.org/10.1145/3340764.3340793

Arun Kulshreshth, Jonas Schild, and Joseph J. LaViola. 2012. Evaluating user performance in 3D stereo and motion enabled video games. *Foundations of Digital Games 2012, FDG 2012 – Conference Program* (2012), 33–40. https://doi.org/10.1145/2282338.2282350

Andrew L. Kun. 2018. Human-Machine Interaction for Vehicles: Review and Outlook. *Foundations and Trends in Human–Computer Interaction* 11, 4 (2018), 201–293. https://doi.org/10.1561/1100000069.Andrew

Andrew L Kun, Susanne Boll, and Albrecht Schmidt. 2016. Shifting Gears: User Interfaces in the Age of Autonomous Driving. *IEEE Pervasive Computing* 15, 1 (jan 2016), 32–38. https://doi.org/10.1109/MPRV.2016.14

Simon M. Laham, Peter Koval, and Adam L. Alter. 2012. The name-pronunciation effect: Why people like Mr. Smith more than Mr. Colquhoun. *Journal of Experimental Social Psychology* 48, 3 (may 2012), 752–756. https://doi.org/10.1016/j.jesp.2011.12.002

Matthew Lakier, Lennart E Nacke, Takeo Igarashi, and Daniel Vogel. 2019. Cross-Car, Multiplayer Games for Semi-Autonomous Driving. In *Proceedings of the Annual Symposium on Computer-Human Interaction in Play.* ACM, New York, NY, USA, 467–480. https://doi.org/10.1145/3311350.3347166

Marc Lambooij, Wijnand IJsselsteijn, Marten Fortuin, and Ingrid Heynderickx. 2009. Visual Discomfort and Visual Fatigue of Stereoscopic Displays: A Review. *Journal of Imaging Science and Technology* 53, 3 (2009), 030201. https://doi.org/10.2352/J.ImagingSci.Technol.2009.53.3.030201

Sabine Langlois and Boussaad Soualmi. 2016. Augmented reality versus classical HUD to take over from automated driving: An aid to smooth reactions and to anticipate maneuvers. In *2016 IEEE 19th International Conference on Intelligent Transportation Systems (ITSC).* IEEE, 1571–1578. https://doi.org/10.1109/ITSC.2016.7795767

David R. Large, Gary Burnett, Andrew Morris, Arun Muthumani, and Rebecca Matthias. 2018. A Longitudinal Simulator Study to Explore Drivers' Behaviour During Highly-Automated Driving. In *Advances in Intelligent Systems and Computing*, Neville A Stanton (Ed.). Advances in Intelligent Systems and Computing, Vol. 597. Springer International Publishing, Cham, 583–594. https://doi.org/10.1007/978-3-319-60441-1_57

David R. Large, Gary Burnett, Davide Salanitri, Anneka Lawson, and Elizabeth Box. 2019. A Longitudinal Simulator Study to Explore Drivers' Behaviour in Level 3 Automated Vehicles. In *Proceedings of the 11th International Conference on Automotive User Interfaces and Interactive Vehicular Applications – AutomotiveUI '19.* ACM Press, New York, New York, USA, 222–232. https://doi.org/10.1145/3342197.3344519

David R. Large, Elizabeth Crundall, Gary Burnett, Catherine Harvey, and Panos Konstantopoulos. 2016. Driving without wings: The effect of different digital mirror locations on the visual behaviour, performance and opinions of drivers. *Applied Ergonomics* 55 (jul 2016), 138–148. https://doi.org/10.1016/j.apergo.2016.02.003

Bettina Laugwitz, Theo Held, and Martin Schrepp. 2008. Construction and Evaluation of a User Experience Questionnaire. In *Symposium of the Austrian HCI and Usability Engineering Group.* Springer, Berlin, Heidelberg, 63–76. https://doi.org/10.1007/978-3-540-89350-9_6

Joseph J. LaViola Jr., Ernst Kruijff, Doug Bowman, Ivan P. Poupyrev, and Ryan P. McMahan. 2017. *3D User Interfaces: Theory and Practice (second edition).* Addison-Wesley.

John D. Lee and Katrina A. See. 2004. Trust in Automation: Designing for Appropriate Reliance. *Human Factors: The Journal of the Human Factors and Ergonomics Society* 46, 1 (2004), 50–80. https://doi.org/10.1518/hfes.46.1.50.30392

Bridget A. Lewis and Carryl L. Baldwin. 2012. Equating perceived urgency across auditory, visual, and tactile signals. *Proceedings of the Human Factors and Ergonomics Society* (2012), 1307–1311. https://doi.org/10.1177/1071181312561379

Lee Lisle, Coleman Merenda, Kyle Tanous, Hyungil Kim, Joseph L. Gabbard, and Doug A. Bowman. 2019. Effects of Volumetric Augmented Reality Displays on Human Depth Judgments. *International Journal of Mobile Human Computer Interaction* 11, 2 (apr 2019), 1–18. https://doi.org/10.4018/IJMHCI.2019040101

Andreas Löcken, Heiko Muller, Wilko Heuten, and Susanne Boll. 2015. An experiment on ambient light patterns to support lane change decisions. In *2015 IEEE Intelligent Vehicles Symposium (IV).* IEEE, 505–510. https://doi.org/10.1109/IVS.2015.7225735

Benjamin Long, Sue Ann Seah, Tom Carter, and Sriram Subramanian. 2014. Rendering Volumetric Haptic Shapes in Mid-Air using Ultrasound. *ACM Transactions on Graphics* 33 (2014). https://doi.org/10.1145/2661229.2661257

Lutz Lorenz, Philipp Kerschbaum, and Josef Schumann. 2014. Designing take over scenarios for automated driving. *Proceedings of the Human Factors and Ergonomics Society Annual Meeting* 58, 1 (sep 2014), 1681–1685. https://doi.org/10.1177/1541931214581351

Tyron Louw and Natasha Merat. 2017. Are you in the loop? Using gaze dispersion to understand driver visual attention during vehicle automation. *Transportation Research Part C: Emerging Technologies* 76 (2017), 35–50. https://doi.org/10.1016/j.trc.2017.01.001

Ernst Lueder. 2012. *3D Displays*. Wiley, Chichester, West Sussex.

Gerhard Marquart, Christopher Cabrall, and Joost de Winter. 2015. Review of Eye-related Measures of Drivers' Mental Workload. In *Proceedings of the 6th International Conference on Applied Human Factors and Ergonomics (AHFE 2015) and the Affiliated Conferences, AHFE 2015*. The Authors, 2854–2861. https://doi.org/10.1016/j.promfg.2015.07.783

Sandra P. Marshall. 2003. The Index of Cognitive Activity: measuring cognitive workload. In *Proceedings of the IEEE 7th Conference on Human Factors and Power Plants*. IEEE, 7–5–7–9. https://doi.org/10.1109/HFPP.2002.1042860

Kohei Matsumura and David S Kirk. 2018. On Active Passengering: Supporting In-Car Experiences. In *Proceedings of the ACM on Interactive, Mobile, Wearable and Ubiquitous Technologies*. 1–23. https://doi.org/10.1145/3161176

Stefan Mattes. 2003. The Lane Change Task as a Tool For Driver Distraction Evaluation. *Quality of work and products in enterprises of the future* (2003), 1–30. http://www-nrd.nhtsa.dot.gov/departments/nrd-01/IHRA/ITS/MATTES.pdf

John W. Mauchly. 1940. Significance test for sphericity of a normal n-variate distribution. *The Annals of Mathematical Statistics* 11, 2 (1940), 204–209.

Keenan R. May, Thomas M. Gable, and Bruce N. Walker. 2017. Designing an In-Vehicle Air Gesture Set Using Elicitation Methods. *Proceedings of the 9th International Conference on Automotive User Interfaces and Interactive Vehicular Applications – AutomotiveUI '17* (2017), 74–83. https://doi.org/10.1145/3122986.3123015

Roderick McCall, Fintan McGee, Alexander Meschtscherjakov, Nicolas Louveton, and Thomas Engel. 2016. Towards A Taxonomy of Autonomous Vehicle Handover Situations. *Proceedings of the 8th International Conference on Automotive User Interfaces and Interactive Vehicular Applications (Automotive'UI 16)* (2016), 193–200. https://doi.org/10.1145/3003715.3005456

Louis McCallum and Peter W. McOwan. 2014. Shut up and play: A musical approach to engagement and social presence in Human Robot Interaction. In *The 23rd IEEE International Symposium on Robot and Human Interactive Communication*. IEEE, 949–954. https://doi.org/10.1109/ROMAN.2014.6926375

John P. McIntire, Paul R. Havig, and Eric E. Geiselman. 2012. What is 3D good for? A review of human performance on stereoscopic 3D displays. In *Proc of SPIE*, Vol. 8383. 83830X. https://doi.org/10.1117/12.920017

John P. McIntire, Paul R. Havig, and Eric E. Geiselman. 2014. Stereoscopic 3D displays and human performance: A comprehensive review. *Displays* 35, 1 (jan 2014), 18–26. https://doi.org/10.1016/j.displa.2013.10.004

John P. McIntire, Paul R. Havig, and Alan R. Pinkus. 2015. A guide for human factors research with stereoscopic 3D displays. *SPIE Defense + Security* 9470 (2015), 94700A. https://doi.org/10.1117/12.2176997

John P. McIntire and Kristen K. Liggett. 2015. The (possible) utility of stereoscopic 3D displays for information visualization: The good, the bad, and the ugly. *2014 IEEE VIS International Workshop on 3DVis, 3DVis 2014* (2015), 1–9. https://doi.org/10.1109/3DVis.2014.7160093

Ryan D. McKendrick and Erin Cherry. 2018. A Deeper Look at the NASA TLX and Where It Falls Short. *Proceedings of the Human Factors and Ergonomics Society Annual Meeting* 62, 1 (sep 2018), 44–48. https://doi.org/10.1177/1541931218621010

Michael McKenna. 1992. Interactive viewpoint control and three-dimensional operations. *Proceedings of the 1992 symposium on Interactive 3D graphics – I3D '92* (1992), 53–56. https://doi.org/10.1145/147156.147163

Vivien Melcher, Stefan Rauh, Frederik Diederichs, Harald Widlroither, and Wilhelm Bauer. 2015. Take-Over Requests for Automated Driving. *Procedia Manufacturing* 3, Ahfe (2015), 2867–2873. https://doi.org/10.1016/j.promfg.2015.07.788

Coleman Merenda, Chihiro Suga, Joseph Gabbard, and Teruhisa Misu. 2019. Effects of Vehicle Simulation Visual Fidelity on Assessing Driver Performance and Behavior. *2019 IEEE Intelligent Vehicles Symposium (IV)* (2019), 1679–1686. https://doi.org/10.1109/ivs.2019.8813863

Alexander Meschtscherjakov, Wendy Ju, Manfred Tscheligi, Dalila Szostak, Sven Krome, Bastian Pfleging, Rabindra Ratan, Ioannis Politis, Sonia Baltodano, and Dave Miller. 2016. HCI and Autonomous Vehicles. In *Proceedings of the 2016 CHI Conference Extended Abstracts on Human Factors in Computing Systems – CHI EA '16*. ACM Press, New York, New York, USA, 3542–3549. https://doi.org/10.1145/2851581.2856489

Dave Miller, Annabel Sun, and Wendy Ju. 2014. Situation awareness with different levels of automation. In *Conference Proceedings – IEEE International Conference on Systems, Man and Cybernetics*. 688–693. https://doi.org/10.1109/SMC.2014.6973989

Brian Mok, Mishel Johns, David Miller, and Wendy Ju. 2017. Tunneled In: Drivers with Active Secondary Tasks Need More Time to Transition from Automation. *Proceedings of the 2017 CHI Conference on Human Factors in Computing Systems – CHI '17* (2017), 2840–2844. https://doi.org/10.1145/3025453.3025713

Henry Moller. 2011. Psychatric Disorders and Driving Performance. In *Handbook of Driving Simulation for Engineering, Medicine, and Psychology* (1st ed.), Donald L. Fisher, Matthew Rizzo, Jeff K. Caird, and John D. Lee (Eds.). CRC Press, Boca Raton, FL, Chapter 47, 672–685.

Masahiro Mori, Karl F. MacDorman, and Norri Kageki. 2012. The uncanny valley. *IEEE Robotics and Automation Magazine* 19, 2 (2012), 98–100. https://doi.org/10.1109/MRA.2012.2192811

Meredith Ringel Morris, Andreea Danielescu, Steven Drucker, Danyel Fisher, Bongshin Lee, C. Schraefel, and Jacob O. Wobbrock. 2014. Reducing legacy bias in gesture elicitation studies. *Interactions* 21, 3 (2014), 40–45. https://doi.org/10.1145/2591689 arxiv:arXiv:1011.1669v3

M@RS – The Digital Archives of Mercedes-Benz. 2019. 169 series A-Class Coupe, 2004 – 2008. https://mercedes-benz-publicarchive.com/marsClassic/en/instance/ko.xhtml?oid=453317

Lothar Muhlbach, Martin Bocker, and Angela Prussog. 1995. Telepresence in Videocommunications: A Study on Stereoscopy and Individual Eye Contact. *Human Factors: The Journal of the Human Factors and Ergonomics Society* 37, 2 (jun 1995), 290–305. https://doi.org/10.1518/001872095779064582

Bonne M. Muir. 1994. Trust in automation: Part I. Theoretical issues in the study of trust and human intervention in automated systems. *Ergonomics* 37, 11 (1994), 1905–1922. https://doi.org/10.1080/00140139408964957

Nadia Mullen, Judith Charlton, Anna Devlin, and Michel Bedard. 2011. Simulator Validty: Behaviors Observed on the Simulator and on the Road. In *Handbook of Driving Simulation for Engineering, Medicine, and Psychology* (1st ed.), Donald L. Fisher, Matthew Rizzo, Jeff K. Caird, and John D. Lee (Eds.). CRC Press, Boca Raton, FL, Boca Raton, FL, Chapter 13, 198–214.

Nermine Munla, Mohamad Khalil, Ahmad Shahin, and Azzam Mourad. 2015. Driver stress level detection using HRV analysis. *2015 International Conference on Advances in Biomedical Engineering, ICABME 2015* (2015), 61–64. https://doi.org/10.1109/ICABME.2015.7323251

Miguel A. Nacenta, Satoshi Sakurai, Tokuo Yamaguchi, Yohei Miki, Yuichi Itoh, Yoshifumi Kitamura, Sriram Subramanian, and Carl Gutwin. 2007. E-conic: a Perspective-Aware Interface for Multi-Display Environments. In *Proceedings of the 20th annual ACM symposium on User interface software and technology – UIST '07*. ACM Press, New York, New York, USA, 279. https://doi.org/10.1145/1294211.1294260

Neelam Naikar. 1998. *Perspective Displays: A Review of Human Factors*. Technical Report. Defence Science And Technology Organisation Canberra (Australia). https://catalogue.nla.gov.au/Record/2840016

Clifford Nass and Youngme Moon. 2000. Machines and Mindlessness: Social Responses to Computers. *Journal of Social Issues* 56, 1 (jan 2000), 81–103. https://doi.org/10.1111/0022-4537.00153

Frederik Naujoks, Christoph Mai, and Alexandra Neukum. 2014. The Effect of Urgency of Take-Over Requests During Highly Automated Driving Under Distraction Conditions. *Proceedings of the 5th International Conference on Applied Human Factors and Ergonomics AHFE* (2014), 2099–2106.

Victor Ng-Thow-Hing, Karlin Bark, Lee Beckwith, Cuong Tran, Rishabh Bhandari, and Srinath Sridhar. 2013. User-centered perspectives for automotive augmented reality. In *2013 IEEE International Symposium on Mixed and Augmented Reality – Arts, Media, and Humanities, ISMAR-AMH 2013*. 13–22. https://doi.org/10.1109/ISMAR-AMH.2013.6671262

Kristine L. Nowak and Frank Biocca. 2003. The Effect of the Agency and Anthropomorphism on Users' Sense of Telepresence, Copresence, and Social Presence in Virtual Environments. *Presence: Teleoperators and Virtual Environments* 12, 5 (oct 2003), 481–494. https://doi.org/10.1162/105474603322761289

Eshed Ohn-Bar, Cuong Tran, and Mohan Trivedi. 2012. Hand gesture-based visual user interface for infotainment. In *Proceedings of the 4th International Conference on Automotive User Interfaces and Interactive Vehicular Applications – AutomotiveUI '12*. ACM Press, New York, New York, USA, 111. https://doi.org/10.1145/2390256.2390274

Satoshi Okamoto and Shin Sano. 2017. Anthropomorphic AI Agent Mediated Multimodal Interactions in Vehicles. In *Proceedings of the 9th International Conference on Automotive*

User Interfaces and Interactive Vehicular Applications Adjunct – AutomotiveUI '17. ACM Press, New York, New York, USA, 110–114. https://doi.org/10.1145/3131726.3131736

Luis Oliveira, Jacob Luton, Sumeet Iyer, Chris Burns, Alexandros Mouzakitis, Paul Jennings, and Stewart Birrell. 2018. Evaluating How Interfaces Influence the User Interaction with Fully Autonomous Vehicles. In *Proceedings of the 10th International Conference on Automotive User Interfaces and Interactive Vehicular Applications – AutomotiveUI '18*. ACM Press, New York, New York, USA, 320–331. https://doi.org/10.1145/3239060.3239065

Hector Olmedo, David Escudero, and Valentín Cardeñoso. 2015. Multimodal interaction with virtual worlds XMMVR: eXtensible language for MultiModal interaction with virtual reality worlds. *Journal on Multimodal User Interfaces* 9, 3 (2015), 153–172. https://doi.org/10.1007/s12193-015-0176-5

OpenDS. 2016. OpenDS – The flexible open source driving simulator. https://www.opends.eu/

Oskar Palinko, Andrew L. Kun, Zachary Cook, Adam Downey, Aaron Lecomte, Meredith Swanson, and Tina Tomaszewski. 2013. Towards augmented reality navigation using affordable technology. *Proceedings of the 5th International Conference on Automotive User Interfaces and Interactive Vehicular Applications – AutomotiveUI '13* (2013), 238–241. https://doi.org/10.1145/2516540.2516569

Joanna Paliszkiewicz. 2017. The Foundations of Trust. In *Intuition, Trust, and Analytics* (1 ed.), Jay Liebowitz, Joanna Olga Paliszkiewicz, and Jerzy Gołuchowski (Eds.). Auerbach Publications, Boca Raton, Florida: CRC Press, [2018], 69–82. https://doi.org/10.1201/9781315195551-5

Peter Ludwig Panum. 1858. *Physiologische Untersuchungen ueber das Sehen mit zwei Augen*. Schwers, Kiel. 94 pages. http://books.google.com/books?id=xLErwWTZmR0C

Raja Parasuraman and Christopher A. Miller. 2004. Trust and etiquette in high-criticality automated systems. *Commun. ACM* 47, 4 (apr 2004), 51. https://doi.org/10.1145/975817.975844

Raja Parasuraman and Victor Riley. 1997. Humans and Automation: Use, Misuse, Disuse, Abuse. *Human Factors: The Journal of the Human Factors and Ergonomics Society* 39, 2 (jun 1997), 230–253. https://doi.org/10.1518/001872097778543886

Randy Pausch, Thomas Crea, and Matthew Conway. 1992. A Literature Survey for Virtual Environments: Military Flight Simulator Visual Systems and Simulator Sickness. *Presence: Teleoperators and Virtual Environments* 1, 3 (jan 1992), 344–363. https://doi.org/10.1162/pres.1992.1.3.344

Annie Pauzie. 2008. A method to assess the driver mental workload: The driving activity load index (DALI). *IET Intelligent Transport Systems* 2, 4 (2008), 315. https://doi.org/10.1049/iet-its:20080023

Julie Paxion, Edith Galy, and Catherine Berthelon. 2014. Mental workload and driving. *Frontiers in Psychology* 5 (dec 2014). https://doi.org/10.3389/fpsyg.2014.01344

Jeunese Payne, Andrea Szymkowiak, Paul Robertson, and Graham Johnson. 2013. Gendering the Machine: Preferred Virtual Assistant Gender and Realism in Self-Service. In *Lecture Notes in Computer Science (including subseries Lecture Notes in Artificial Intelligence and Lecture Notes in Bioinformatics)*, R. Aylett, B. Krenn, Pelachaud C., and H. Shimodaira (Eds.). Springer, Berlin, Heidelberg, 106–115. https://doi.org/10.1007/978-3-642-40415-3_9

Daniel Perez-Marcos, Maria V. Sanchez-Vives, and Mel Slater. 2012. Is my hand connected to my body? The impact of body continuity and arm alignment on the virtual hand illusion. *Cognitive Neurodynamics* 6, 4 (2012), 295–305. https://doi.org/10.1007/s11571-011-9178-5

Sebastiaan Petermeijer, Pavlo Bazilinskyy, Klaus Bengler, and Joost de Winter. 2017a. Take-over again: Investigating multimodal and directional TORs to get the driver back into the loop. *Applied Ergonomics* 62 (2017), 204–215. https://doi.org/10.1016/j.apergo.2017.02.023 arxiv:arXiv:1011.1669v3

Sebastiaan. M. Petermeijer, Stephan Cieler, and Joost C.F. de Winter. 2017b. Comparing spatially static and dynamic vibrotactile take-over requests in the driver seat. *Accident Analysis and Prevention* 99 (2017), 218–227. https://doi.org/10.1016/j.aap.2016.12.001

Ryan A. Peterson and Joseph E. Cavanaugh. 2019. Ordered quantile normalization: a semi-parametric transformation built for the cross-validation era. *Journal of Applied Statistics* (jun 2019). https://doi.org/10.1080/02664763.2019.1630372

Peugeot Deutschland GmbH. 2019. Der neue PEUGEOT 208., 5 pages. https://www.peugeot.de/modelle/alle-modelle/neuer-peugeot-208.html

Nicolas Pfeuffer, Alexander Benlian, Henner Gimpel, and Oliver Hinz. 2019. Anthropomorphic Information Systems. *Business & Information Systems Engineering* 61, 4 (aug 2019), 523–533. https://doi.org/10.1007/s12599-019-00599-y

Bastian Pfleging, Maurice Rang, and Nora Broy. 2016. Investigating user needs for non-driving-related activities during automated driving. In *Proceedings of the 15th International Conference on Mobile and Ubiquitous Multimedia – MUM '16*. ACM Press, New York, New York, USA, 91–99. https://doi.org/10.1145/3012709.3012735

Tran Pham, Jo Vermeulen, Anthony Tang, and Lindsay MacDonald Vermeulen. 2018. Scale Impacts Elicited Gestures for Manipulating Holograms. In *Proceedings of the 2018 on Designing Interactive Systems Conference 2018 – DIS '18*. 227–240. https://doi.org/10.1145/3196709.3196719

Matthew J. Pitts, Elvir Hasedžić, Lee Skrypchuk, Alex Attridge, and Mark Williams. 2015. Adding Depth: Establishing 3D Display Fundamentals for Automotive Applications. https://doi.org/10.4271/2015-01-0147

Thammathip Piumsomboon, Adrian Clark, Mark Billinghurst, and Andy Cockburn. 2013a. User-Defined Gestures for Augmented Reality. *Lecture Notes in Computer Science (including subseries Lecture Notes in Artificial Intelligence and Lecture Notes in Bioinformatics)* 8118 LNCS, PART 2 (2013), 282–299. https://doi.org/10.1007/978-3-642-40480-1_18

Thammathip Piumsomboon, Adrian Clark, Mark Billinghurst, and Andy Cockburn. 2013b. User-Defined Gestures for Augmented Reality BT – Human-Computer Interaction – INTERACT 2013. In *Human-Computer Interaction – INTERACT 2013*. 282–299. https://link.springer.com/chapter/10.1007/978-3-642-40480-1_18

Ioannis Politis, Stephen Brewster, and Frank Pollick. 2013. Evaluating multimodal driver displays of varying urgency. In *Proceedings of the 5th International Conference on Automotive User Interfaces and Interactive Vehicular Applications – AutomotiveUI '13*. ACM Press, New York, New York, USA, 92–99. https://doi.org/10.1145/2516540.2516543

Ioannis Politis, Stephen Brewster, and Frank Pollick. 2015. Language-based multimodal displays for the handover of control in autonomous cars. *Proceedings of the 7th International Conference on Automotive User Interfaces and Interactive Vehicular Applications – AutomotiveUI '15* (2015), 3–10. https://doi.org/10.1145/2799250.2799262

Monika Pölönen, Marja Salmimaa, Viljakaisa Aaltonen, Jukka Häkkinen, and Jari Takatalo. 2009. Subjective measures of presence and discomfort in viewers of color-separation-based stereoscopic cinema. *Journal of the Society for Information Display* 17, 5 (2009), 459. https://doi.org/10.1889/JSID17.5.459

Jeffrey Pruetz, Channing Watson, Todd Tousignant, and Kiran Govindswamy. 2019. Assessment of Automotive Environmental Noise on Mobile Phone Hands-Free Call Quality. In *Noise and Vibration Conference & Exhibition.* https://doi.org/10.4271/2019-01-1597

Hang Qiu, Fawad Ahmad, Fan Bai, Marco Gruteser, and Ramesh Govindan. 2018. AVR: Augmented Vehicular Reality. In *Proceedings of the 16th Annual International Conference on Mobile Systems, Applications, and Services.* ACM, New York, NY, USA, 81–95. https://doi.org/10.1145/3210240.3210319

Lingyun Qiu and Izak Benbasat. 2005. Online Consumer Trust and Live Help Interfaces: The Effects of Text-to-Speech Voice and Three-Dimensional Avatars. *International Journal of Human-Computer Interaction* 19, 1 (sep 2005), 75–94. https://doi.org/10.1207/s15327590ijhc1901_6

Jonas Radlmayr, Christian Gold, Lutz Lorenz, Mehdi Farid, and Klaus Bengler. 2014. How Traffic Situations and Non-Driving Related Tasks Affect the Take-Over Quality in Highly Automated Driving. *Proceedings of the Human Factors and Ergonomics Society Annual Meeting* 58, 1 (2014), 2063–2067. https://doi.org/10.1177/1541931214581434

Eric D. Ragan, Regis Kopper, Philip Schuchardt, and Doug A. Bowman. 2013. Studying the Effects of Stereo, Head Tracking, and Field of Regard on a Small-Scale Spatial Judgment Task. *IEEE Transactions on Visualization and Computer Graphics* 19, 5 (may 2013), 886–896. https://doi.org/10.1109/TVCG.2012.163

Juho Rantakari, Jani Väyrynen, Ashley Colley, and Jonna Häkkilä. 2017. Exploring the design of stereoscopic 3D for multilevel maps. In *Proceedings of the 6th ACM International Symposium on Pervasive Displays – PerDis '17.* ACM Press, New York, New York, USA, 1–6. https://doi.org/10.1145/3078810.3078829

Jenny C. A. Read. 2015. What is stereoscopic vision good for?. In *Stereoscopic Displays and Applications XXVI,* Nicolas S. Holliman, Andrew J. Woods, Gregg E. Favalora, and Takashi Kawai (Eds.). 93910N. https://doi.org/10.1117/12.2184988

Miguel Ángel Recarte, Elisa Pérez, Ángela Conchillo, and Luis Miguel Nunes. 2008. Mental Workload and Visual Impairment: Differences between Pupil, Blink, and Subjective Rating. *The Spanish Journal of Psychology* 11, 2 (nov 2008), 374–385. https://doi.org/10.1017/S1138741600004406

Stephan Reichelt, Ralf Haussler, Gerald Fütterer, and Norbert Leister. 2010. Depth cues in human visual perception and their realization in 3D displays. *Three Dimensional Imaging, Visualization, and Display 2010* 7690, 0 (2010), 76900B–76900B–12. https://doi.org/10.1117/12.850094

Gary B. Reid and Thomas E. Nygren. 1988. The Subjective Workload Assessment Technique: A Scaling Procedure for Measuring Mental Workload. In *Advances in Psychology.* 185–218. https://doi.org/10.1016/S0166-4115(08)62387-0

Bryan Reimer, Chuck Gulash, Bruce Mehler, James P. Foley, Stephen Arredondo, and Alexander Waldmann. 2014. The MIT AgeLab n-back. In *Proceedings of the 6th International Conference on Automotive User Interfaces and Interactive Vehicular Applications – AutomotiveUI '14.* ACM Press, New York, New York, USA, 1–6. https://doi.org/10.1145/2667239.2667293

Tara Rezvani, Katherine Driggs-Campbell, Dorsa Sadigh, S. Shankar Sastry, Sanjit A. Seshia, and Ruzena Bajcsy. 2016. Towards trustworthy automation: User interfaces that convey internal and external awareness. In *2016 IEEE 19th International Conference on Intelligent Transportation Systems (ITSC)*. IEEE, 682–688. https://doi.org/10.1109/ITSC.2016.7795627

Andreas Riegler, Andreas Riener, and Clemens Holzmann. 2019. Virtual reality driving simulator for user studies on automated driving. In *Proceedings of the 11th International Conference on Automotive User Interfaces and Interactive Vehicular Applications Adjunct Proceedings – AutomotiveUI '19*. ACM Press, New York, New York, USA, 502–507. https://doi.org/10.1145/3349263.3349595

Andreas Riener, Susanne Boll, and Andrew Kun. 2016. *Automotive User Interfaces in the Age of Automation: report of Dagstuhl Seminar 16262*. Schloss Dagstuhl–Leibniz-Zentrum fuer Informatik. 111–159 pages. https://doi.org/10.4230/DagRep.6.6.111

Andreas Riener, Alois Ferscha, Florian Bachmair, Patrick Hagmüller, Alexander Lemme, Dominik Muttenthaler, David Pühringer, Harald Rogner, Adrian Tappe, and Florian Weger. 2013. Standardization of the in-car gesture interaction space. In *Proceedings of the 5th International Conference on Automotive User Interfaces and Interactive Vehicular Applications (AutomotiveUI '13)*. 14–21. https://doi.org/10.1145/2516540.2516544

Robert Bosch GmbH. 2019. Bosch zeigt 3D-Display fürs Fahrzeug. https://www.bosch-presse.de/pressportal/de/de/neue-dimension-bosch-bringt-3d-display-ins-fahrzeug-196096.html

George Robertson, Maarten van Dantzich, Daniel Robbins, Mary Czerwinski, Ken Hinckley, Kirsten Risden, David Thiel, and Vadim Gorokhovsky. 2000. The Task Gallery. In *Proceedings of the SIGCHI conference on Human factors in computing systems – CHI '00*. ACM Press, New York, New York, USA, 494–501. https://doi.org/10.1145/332040.332482

Florian Roider and Tom Gross. 2018. I See Your Point: Integrating Gaze to Enhance Pointing Gesture Accuracy While Driving. In *Proceedings of the 10th International Conference on Automotive User Interfaces and Interactive Vehicular Applications – AutomotiveUI '18*. ACM Press, New York, New York, USA, 351–358. https://doi.org/10.1145/3239060.3239084

Sonja Rümelin. 2014. *The Cockpit for the 21st century*. Dissertation. Ludwig-Maximilians-Universität München.

SAE International. 2015. *Taxonomy and Definitions for Terms Related to Driving Automation Systems for On-Road Motor Vehicles (SAE J 3016)*. Technical Report. SAE International. http://standards.sae.org/j3016_201609/https://www.sae.org/standards/content/j3016_201806/

SAE On-Road Automated Vehicle Standards Committee. 2014. Taxonomy and definitions for terms related to on-road motor vehicle automated driving systems. *SAE Standard J3016* (2014). http://www.sae.org/misc/pdfs/automated_driving.pdf

Stefan Schaal. 2019. Historical Perspective of Humanoid Robot Research in the Americas. In *Humanoid Robotics: A Reference*. Springer Netherlands, Dordrecht, 9–17. https://doi.org/10.1007/978-94-007-6046-2_143

Clemens Schartmüller, Andreas Riener, Philipp Wintersberger, and Anna-Katharina Frison. 2018. Workaholistic: On Balancing Typing- and Handover-Performance in Automated Driving. *Proceedings of the 20th International Conference on Human-Computer Inter-*

action with Mobile Devices and Services, MobileHCI'18 (2018), 16:1–16:12. https://doi. org/10.1145/3229434.3229459

Jonas Schild, Joseph LaViola, and Maic Masuch. 2012. Understanding user experience in stereoscopic 3D games. In *Proceedings of the 2012 ACM annual conference on Human Factors in Computing Systems – CHI '12.* ACM Press, New York, New York, USA, 89. https://doi.org/10.1145/2207676.2207690

Martin Schrepp, Andreas Hinderks, and Joerg Thomaschewski. 2017. Construction of a Benchmark for the User Experience Questionnaire (UEQ). *International Journal of Interactive Multimedia and Artificial Intelligence* 4, 4 (2017), 40. https://doi.org/10.9781/ijimai. 2017.445

Katie E. Schwab, Nathan J. Curtis, Martin B. Whyte, Ralph V. Smith, Timothy A. Rockall, Karen Ballard, and Iain C. Jourdan. 2019. 3D laparoscopy does not reduce operative duration or errors in day-case laparoscopic cholecystectomy: a randomised controlled trial. *Surgical Endoscopy* (2019). https://doi.org/10.1007/s00464-019-06961-1

Bennett L. Schwartz and John H. Krantz. 2018. Random Dot Stereograms. https://isle.hanover. edu/Ch07DepthSize/Ch07RandomDotStereograms.html

Felix Schwarz and Wolfgang Fastenmeier. 2018. Visual advisory warnings about hidden dangers: Effects of specific symbols and spatial referencing on necessary and unnecessary warnings. *Applied Ergonomics* 72, May (oct 2018), 25–36. https://doi.org/10.1016/j. apergo.2018.05.001

Bobbie D. Seppelt and John D. Lee. 2007. Making adaptive cruise control (ACC) limits visible. *International Journal of Human-Computer Studies* 65, 3 (mar 2007), 192–205. https://doi. org/10.1016/j.ijhcs.2006.10.001

Jun'ichiro Seyama and Ruth S. Nagayama. 2007. The Uncanny Valley: Effect of Realism on the Impression of Artificial Human Faces. *Presence: Teleoperators and Virtual Environments* 16, 4 (aug 2007), 337–351. https://doi.org/10.1162/pres.16.4.337

Gözel Shakeri, John H. Williamson, and Stephen Brewster. 2018. May the Force Be with You: Ultrasound Haptic Feedback for Mid-Air Gesture Interaction in Cars. In *Proceedings of the 10th International Conference on Automotive User Interfaces and Interactive Vehicular Applications – AutomotiveUI '18.* ACM Press, New York, New York, USA, 1–10. https:// doi.org/10.1145/3239060.3239081

Takashi Shibata, Joohwan Kim, David M. Hoffman, and Martin S. Banks. 2011. The zone of comfort: Predicting visual discomfort with stereo displays. *Journal of vision* 11, 8 (2011), 1–29. https://doi.org/10.1167/11.8.11.Introduction arxiv: NIHMS150003

Chaklam Silpasuwanchai and Xiangshi Ren. 2015. Designing concurrent full-body gestures for intense gameplay. *International Journal of Human Computer Studies* 80 (2015), 1–13. https://doi.org/10.1016/j.ijhcs.2015.02.010

Sarah M. Simmons, Jeff K. Caird, and Piers Steel. 2017. A meta-analysis of in-vehicle and nomadic voice-recognition system interaction and driving performance. *Accident Analysis & Prevention* 106 (sep 2017), 31–43. https://doi.org/10.1016/j.aap.2017.05.013

Paul Slovic. 1993. Perceived Risk, Trust, and Democracy. *Risk Analysis* 13, 6 (1993), 675–682. https://doi.org/10.1111/j.1539-6924.1993.tb01329.x

Andreas Sonderegger and Juergen Sauer. 2009. The influence of laboratory set-up in usability tests: effects on user performance, subjective ratings and physiological measures. *Ergonomics* 52, 11 (nov 2009), 1350–1361. https://doi.org/10.1080/00140130903067797

Ani Sreedhar and Ashok Menon. 2019. Understanding and evaluating diplopia. *Kerala Journal of Ophthalmology* 31, 2 (2019), 102. https://doi.org/10.4103/kjo.kjo_57_19

Kay Stanney, Robert Kennedy, Deborah Harm, Daniel Compton, D Lanham, and Julie Drexler. 2010. *Configural Scoring of Simulator Sickness, Cybersickness and Space Adaptation Syndrome.* Technical Report. NASA Johnson Space Center. 247–278 pages. https://doi.org/10.1201/9781410608888.ch12

Kay M. Stanney and Jacob O. Wobbrock. 2019. ARTool: Aligned Rank Transform for Nonparametric Factorial ANOVAs. R package version 0.10.6. https://doi.org/10.5281/zenodo.594511

John A. Stern and June J. Skelly. 1984. The Eye Blink and Workload Considerations. *Proceedings of the Human Factors Society Annual Meeting* 28, 11 (oct 1984), 942–944. https://doi.org/10.1177/154193128402801101

Heather A. Stoner, Donald L. Fisher, and Michael Mollenhauser Jr. 2011. Simulator And Scenario Factors Influencing Simulator Sickness. In *2nd Handbook of Driving Simulation for Engineering, Medicine, and Psychology* (1st ed.), Donald L. Fisher, Matthew Rizzo, Jeff K. Caird, and John D. Lee (Eds.). CRC Press, Boca Raton, FL, Chapter 14, 217–240.

Richard Swette, Keenan R May, Thomas M Gable, and Bruce N Walker. 2013. Comparing three novel multimodal touch interfaces for infotainment menus. In *Proceedings of the 5th International Conference on Automotive User Interfaces and Interactive Vehicular Applications – AutomotiveUI '13.* ACM Press, New York, New York, USA, 100–107. https://doi.org/10.1145/2516540.2516559

Joseph Szczerba and Roger Hersberger. 2014. The Use of Stereoscopic Depth in an Automotive Instrument Display: Evaluating User-Performance in Visual Search and Change Detection. *Proceedings of the Human Factors and Ergonomics Society Annual Meeting* 58, 1 (sep 2014), 1184–1188. https://doi.org/10.1177/1541931214581247

Jari Takatalo, Takashi Kawai, Jyrki Kaistinen, Göte Nyman, and Jukka Häkkinen. 2011. User Experience in 3D Stereoscopic Games. *Media Psychology* 14, 4 (nov 2011), 387–414. https://doi.org/10.1080/15213269.2011.620538

Wa James Tam, Andre Vincent, Ronald Renaud, Phil Blanchfield, and Taali Martin. 2003. Comparison of stereoscopic and nonstereoscopic video images for visual telephone systems. *Stereoscopic Displays and Virtual Reality Systems X* 5006, May 2003 (2003), 304. https://doi.org/10.1117/12.474108

Alexander-Cosmin Teleki, Maria Fritz, and Matthias Kreimeyer. 2017. Use cases for automated driving commercial vehicles. In *Fahrerassistenzsysteme 2017. Proceedings.* Springer Vieweg, Wiesbaden, 187–200. https://doi.org/10.1007/978-3-658-19059-0_11

Ariel Telpaz, Brian Rhindress, Ido Zelman, and Omer Tsimhoni. 2015. Haptic seat for automated driving. In *Proceedings of the 7th International Conference on Automotive User Interfaces and Interactive Vehicular Applications – AutomotiveUI '15.* ACM Press, New York, New York, USA, 23–30. https://doi.org/10.1145/2799250.2799267

Marcus Tönnis, Verena Broy, and Gudrun Klinker. 2006. A Survey of Challenges Related to the Design of 3D User Interfaces for Car Drivers. In *3D User Interfaces (3DUI'06),* Vol. 2006. IEEE, 127–134. https://doi.org/10.1109/VR.2006.19

Theophanis Tsandilas. 2018. Fallacies of Agreement: A Critical Review of Consensus Assessment Methods for Gesture Elicitation. *ACM Transactions on Computer-Human Interaction* 25, 3 (2018), 1–49. https://doi.org/10.1145/3182168>

Kaisa Väänänen-Vainio-Mattila, Jani Heikkinen, Ahmed Farooq, Grigori Evreinov, Erno Mäkinen, and Roope Raisamo. 2014. User experience and expectations of haptic feedback in in-car interaction. In *Proceedings of the 13th International Conference on Mobile and Ubiquitous Multimedia – MUM '14*. 248–251. https://doi.org/10.1145/2677972.2677996

Radu-Daniel Vatavu and Jacob O. Wobbrock. 2015. Formalizing Agreement Analysis for Elicitation Studies. In *Proceedings of the 33rd Annual ACM Conference on Human Factors in Computing Systems – CHI '15*. ACM Press, New York, New York, USA, 1325–1334. https://doi.org/10.1145/2702123.2702223

Leena Ventä-Olkkonen, Maaret Posti, and Jonna Häkkilä. 2013. How to Use 3D in Stereoscopic Mobile User Interfaces: Study on Initial User Perceptions. In *Proceedings of International Conference on Making Sense of Converging Media (AcademicMindTrek '13)*. ACM, New York, NY, USA, 39:39–39:42. https://doi.org/10.1145/2523429.2523447

Frank M. F. Verberne. 2015. *Trusting a virtual driver: similarity as a trust cue*. Dissertation. Technische Universiteit Eindhoven. https://research.tue.nl/en/publications/trusting-a-virtual-driver-similarity-as-a-trust-cue

Tobias Vogelpohl, Matthias Kühn, Thomas Hummel, Tina Gehlert, and Mark Vollrath. 2018. Transitioning to manual driving requires additional time after automation deactivation. *Transportation Research Part F: Traffic Psychology and Behaviour* 55 (2018), 464–482. https://doi.org/10.1016/j.trf.2018.03.019

Panagiotis Vogiatzidakis and Panayiotis Koutsabasis. 2018. Gesture Elicitation Studies for Mid-Air Interaction: A Review. *Multimodal Technologies and Interaction* 2, 4 (sep 2018), 65. https://doi.org/10.3390/mti2040065

Tamara von Sawitzky, Philipp Wintersberger, Andreas Riener, and Joseph L. Gabbard. 2019. Increasing trust in fully automated driving: route indication on an augmented reality headup display. In *Proceedings of the 8th ACM International Symposium on Pervasive Displays*. ACM, New York, NY, USA, 1–7. https://doi.org/10.1145/3321335.3324947

Marcel Walch, Kristin Lange, Martin Baumann, and Michael Weber. 2015. Autonomous Driving: Investigating the Feasibility of Car-driver Handover Assistance. *Proceedings of the 7th International Conference on Automotive User Interfaces and Interactive Vehicular Applications* (2015), 11–18. https://doi.org/10.1145/2799250.2799268

Ping Wan, Chaozhong Wu, Yingzi Lin, and Xiaofeng Ma. 2019. Driving Anger States Detection Based on Incremental Association Markov Blanket and Least Square Support Vector Machine. *Discrete Dynamics in Nature and Society* 2019 (2019). https://doi.org/10.1155/2019/2745381

Lili Wang, Kees Teunissen, Yan Tu, Li Chen, Panpan Zhang, Tingting Zhang, and Ingrid Heynderickx. 2011. Crosstalk evaluation in stereoscopic displays. *IEEE/OSA Journal of Display Technology* 7, 4 (2011), 208–214. https://doi.org/10.1109/JDT.2011.2106760

Ying Wang, Bruce Mehler, Bryan Reimer, Vincent Lammers, Lisa A. D'Ambrosio, and Joseph F. Coughlin. 2010. The validity of driving simulation for assessing differences between in-vehicle informational interfaces: A comparison with field testing. *Ergonomics* 53, 3 (2010), 404–420. https://doi.org/10.1080/00140130903464358

Colin Ware. 2020. *Information Visualization* (4th editio ed.). Morgan Kaufmann, Amsterdam.

Adam Waytz, Joy Heafner, and Nicholas Epley. 2014. The mind in the machine: Anthropomorphism increases trust in an autonomous vehicle. *Journal of Experimental Social Psychology* 52 (2014), 113–117. https://doi.org/10.1016/j.jesp.2014.01.005

Thomas Wilhelm Weißgerber. 2019. *Automatisiertes Fahren mit kontaktanalogem Head-Up Display*. Ph.D. Dissertation. Technische Universität München. http://mediatum.ub.tum.de/? id=1435352

Christopher D. Wickens. 2002. Multiple resources and performance prediction. *Theoretical Issues in Ergonomics Science* 3, 2 (jan 2002), 159–177. https://doi.org/10.1080/ 14639220210123806

Gesa Wiegand, Christian Mai, Kai Holländer, and Heinrich Hussmann. 2019. IncarAR: A Design Space Towards 3D Augmented Reality Applications in Vehicles. In *Proceedings of the 11th International Conference on Automotive User Interfaces and Interactive Vehicular Applications – AutomotiveUI '19*. ACM Press, New York, New York, USA, 1–13. https:// doi.org/10.1145/3342197.3344539

Gesa Wiegand, Christian Mai, Yuanting Liu, and Heinrich Hußmann. 2018. Early Take-Over Preparation in Stereoscopic 3D. In *Proceedings of the 10th International Conference on Automotive User Interfaces and Interactive Vehicular Applications – AutomotiveUI '18*. ACM Press, New York, New York, USA, 142–146. https://doi.org/10.1145/3239092. 3265957

Steven P. Williams and Russell V. Parrish. 1990. New computational control techniques and increased understanding for stereo 3-D displays. *Stereoscopic Displays and Applications* 1256, September 1990 (1990), 73. https://doi.org/10.1117/12.19891

Sonja Windhager, Dennis E. Slice, Katrin Schaefer, Elisabeth Oberzaucher, Truls Thorstensen, and Karl Grammer. 2008. Face to Face. *Human Nature* 19, 4 (dec 2008), 331–346. https:// doi.org/10.1007/s12110-008-9047-z

Philipp Wintersberger, Anna-Katharina Frison, Andreas Riener, and Tamara von Sawitzky. 2019. Fostering User Acceptance and Trust in Fully Automated Vehicles: Evaluating the Potential of Augmented Reality. *PRESENCE: Virtual and Augmented Reality* 27, 1 (mar 2019), 46–62. https://doi.org/10.1162/pres_a_00320

Philipp Wintersberger, Andreas Riener, and Anna-Katharina Frison. 2016. Automated Driving System, Male, or Female Driver: Who'd You Prefer? Comparative Analysis of Passengers' Mental Conditions, Emotional States & Qualitative Feedback. In *Proceedings of the 8th International Conference on Automotive User Interfaces and Interactive Vehicular Applications – Automotive'UI 16*. ACM Press, New York, New York, USA, 51–58. https://doi. org/10.1145/3003715.3005410

Philipp Wintersberger, Tamara von Sawitzky, Anna-Katharina Frison, and Andreas Riener. 2017. Traffic Augmentation as a Means to Increase Trust in Automated Driving Systems. In *Proceedings of the 12th Biannual Conference on Italian SIGCHI Chapter – CHItaly '17*. ACM Press, New York, New York, USA, 1–7. https://doi.org/10.1145/3125571.3125600

Jacob O. Wobbrock, Htet Htet Aung, Brandon Rothrock, and Brad A. Myers. 2005. Maximizing the guessability of symbolic input. In *CHI '05 extended abstracts on Human factors in computing systems – CHI '05*. ACM Press, New York, New York, USA, 1869. https:// doi.org/10.1145/1056808.1057043

Jacob O. Wobbrock, Leah Findlater, Darren Gergle, and James J. Higgins. 2011. The aligned rank transform for nonparametric factorial analyses using only anova procedures. In *Proceedings of the 2011 annual conference on Human factors in computing systems – CHI '11*. ACM Press, New York, New York, USA, 143. https://doi.org/10.1145/1978942.1978963

Jacob O. Wobbrock, Meredith Ringel Morris, and Andrew D. Wilson. 2009. User-defined gestures for surface computing. In *Proceedings of the 27th international conference on*

Human factors in computing systems – CHI 09, Vol. 8. ACM Press, New York, New York, USA, 1083. https://doi.org/10.1145/1518701.1518866

Andrew J. Woods and Adin Sehic. 2009. The compatibility of LCD TVs with time-sequential stereoscopic 3D visualization. *Stereoscopic Displays and Applications XX* 7237, February 2009 (2009), 72370N. https://doi.org/10.1117/12.811838

Felix Wulf, Maria Rimini-Doring, Marc Arnon, and Frank Gauterin. 2015. Recommendations Supporting Situation Awareness in Partially Automated Driver Assistance Systems. *IEEE Transactions on Intelligent Transportation Systems* 16, 4 (aug 2015), 2290–2296. https://doi.org/10.1109/TITS.2014.2376572

Shun-nan Yang, Tawny Schlieski, Brent Selmins, Scott C. Cooper, Rina a. Doherty, Philip J. Corriveau, and James E. Sheedy. 2012. Stereoscopic Viewing and Reported Perceived Immersion and Symptoms. *Optometry and Vision Science* 89, 7 (2012), 1068–1080. https://doi.org/10.1097/OPX.0b013e31825da430

B. W. Yap and C. H. Sim. 2011. Comparisons of various types of normality tests. *Journal of Statistical Computation and Simulation* 81, 12 (dec 2011), 2141–2155. https://doi.org/10.1080/00949655.2010.520163

Nick Yee, Jeremy N Bailenson, and Kathryn Rickertsen. 2007. A meta-analysis of the impact of the inclusion and realism of human-like faces on user experiences in interfaces. In *Proceedings of the SIGCHI Conference on Human Factors in Computing Systems – CHI '07*. ACM Press, New York, New York, USA, 1–10. https://doi.org/10.1145/1240624.1240626

In-Kwon Yeo and Richard A Johnson. 2000. A New Family of Power Transformations to Improve Normality or Symmetry. *biometrika Biometrika* 87, 4 (2000), 954–959.

Rami Zarife. 2014. *Integrative Warning Concept for Multiple Driver Assistance Systems*. Dissertation. Julius-Maximilians-Universität Würzburg. https://nbn-resolving.org/urn:nbn:de:bvb:20-opus-101118

Kathrin Zeeb, Axel Buchner, and Michael Schrauf. 2016. Is take-over time all that matters? the impact of visual-cognitive load on driver take-over quality after conditionally automated driving. *Accident Analysis and Prevention* 92 (2016), 230–239. https://doi.org/10.1016/j.aap.2016.04.002

Gao Zhenhai, Le Dinhdat, Hu Hongyu, Yu Ziwen, and Wu Xinyu. 2017. Driver drowsiness detection based on time series analysis of steering wheel angular velocity. *Proceedings – 9th International Conference on Measuring Technology and Mechatronics Automation, ICMTMA 2017* (2017), 99–101. https://doi.org/10.1109/ICMTMA.2017.0031

Jakub Złotowski, Diane Proudfoot, Kumar Yogeeswaran, and Christoph Bartneck. 2015. Anthropomorphism: Opportunities and Challenges in Human–Robot Interaction. *International Journal of Social Robotics* 7, 3 (jun 2015), 347–360. https://doi.org/10.1007/s12369-014-0267-6

Printed in the United States
by Baker & Taylor Publisher Services